The Spatial and Temporal Dynamics of Host–Parasitoid Interactions

Michael P. Hassell

Department of Biology, Imperial College,
Silwood Park

OXFORD

UNIVERSITY PRESS

OXFORD

UNIVERSITY PRESS

Great Clarendon Street, Oxford OX2 6DP

Oxford University Press is a department of the University of Oxford
It furthers the University's aim of excellence in research, scholarship,
and education by publishing worldwide in

Oxford New York

Athens Auckland Bangkok Bogotá Buenos Aires Calcutta
Cape Town Chennai Dar es Salaam Delhi Florence Hong Kong Istanbul
Karachi Kuala Lumpur Madrid Melbourne Mexico City Mumbai
Nairobi Paris São Paulo Singapore Taipei Tokyo Toronto Warsaw

with associated companies in Berlin Ibadan

Oxford is a registered trade mark of Oxford University Press
in the UK and in certain other countries

Published in the United States
by Oxford University Press Inc., New York

© Michael P. Hassell, 2000

The moral rights of the author have been asserted
Database right Oxford University Press (maker)

First published 2000

A catalogue record for this book is available from the British Library
Library of Congress Cataloging in Publication Data
(Data applied for)
ISBN 0 19 854089 2 (Hbk)
ISBN 0 19 854088 4 (Pbk)
Typeset by Footnote Graphics, Warminster, Wilts

Printed in Great Britain
on acid-free paper by
T.J. International Ltd, Padstow

Preface

This book examines in some depth our current understanding of the population dynamics of one kind of interaction—between insect parasitoids and their hosts. Parasitoids are amongst the most abundant of all animals, making up about 10% or more of metazoan species, and very few insect species escape their attack. These interactions were first modelled over 50 years ago, but for many years there was little, good empirical information on the important factors affecting host and parasitoid populations. The models were very simple and their predictions rather divorced from what we see in the field. Nowadays, much more data is available on many components of host–parasitoid systems, from field observations and from laboratory and field experiments, and this allows a much closer interaction between models and data. The result is a body of theory that makes direct contact with real systems in the field, and illustrates how ecological studies can now be advanced in a way that was not possible when theory and data-gathering were treated as rather separate exercises. These are stirring times for host–parasitoid workers who are on the verge of a detailed understanding of what underpins a whole area of population dynamics.

The contents of this book follow on directly from an earlier monograph (Hassell 1978), which showed how many of the basic components of arthropod predator–prey systems could be simply modelled to reveal their effects on the dynamics of the populations. This body of work is reviewed and brought up to date in Chapters 2 and 3 and provides the foundation for the major developments in the subject over the past 20 years that are elaborated in the subsequent chapters. Foremost amongst these have been the exploration of how spatial patchiness and other forms of heterogeneity may affect the population dynamics of two-species systems, the development of a broad-ranging theory for the interaction of more complex host–parasitoid systems involving several species, and the incorporation of age-structure into a wide range of host–parasitoid models. Of these, the study of spatial processes in particular is opening up major challenges in ecology, particularly because of:

(1) the problems of spatial scale that pervade ecology;
(2) the difficulties in disentangling patterns of dynamics that are driven by the interactions themselves from those imposed by the environment; and

(3) the challenge of carrying out field work at spatial scales far larger than the ecologist conventionally works at.

I am deeply indebted to several people for their guidance and help over the years. George Salt first kindled my interest in parasitoids, George Varley and George Gradwell introduced me to the blend of theory and empiricism that characterises much of the work on host–parasitoid population dynamics, Dick Southwood has been a most encouraging mentor, Bob May has always been an inspiration and Mike Bonsall, Mick Crawley, Charles Godfray, Ilkka Hanski, Bob Holt, John Lawton, Steve Pacala, Jeff Waage and Howard Wilson have been excellent foils for discussing a wide range of ecological issues. I am also very grateful to Mike Bonsall, Mick Crawley, Charles Godfray, Bob May, Paul Harvey and Vicky Taylor for reading through the manuscript and offering plentiful and excellent advice.

Silwood Park   M.P.H.
August 1999

Contents

1

Introduction

For much of this century, insect ecologists have questioned what determines the abundance and patterns of fluctuation of insect populations. In a prescient study on the gypsy moth and brown tailed moth, Howard and Fiske (1911) made the first clear statement on the types of factors affecting the population dynamics of insects. Only one kind of mortality factor, their *facultative agencies*, was thought to maintain a 'natural balance' (i.e. equilibrium) by causing 'the destruction of a greater proportionate number of individuals as the insect in question increases in abundance'. Facultative agencies have since been rechristened as *density-dependent* factors by Smith (1935) and are at the heart of all modern ideas about population regulation. Howard and Fiske included insect parasitoids in this category, which they thought important agents regulating insect populations, both in natural and agricultural habitats. Generalist predators, on the other hand, they thought would act 'in a manner which is opposite of facultative' (i.e. *inversely density-dependent*) since 'they are not directly affected by the abundance or scarcity of any single item in their varied menu ... they average to destroy a certain gross number of individuals each year'. Nearly 90 years on from Howard and Fiske, and this rigid categorisation of how natural enemies act no longer applies. For instance, generalist predators *may* act in an inverse density-dependent way, but equally well could be density-dependent or density-independent factors instead. Parasitoids *may* act in a direct density-dependent way, but can also be inversely density-dependent, delayed density-dependent or density-independent, depending on the conditions and what is being measured. There is, therefore, no simple and tidy recipe for the dynamical effects of parasitoids and other natural enemies on their hosts and prey. Their impact is contingent on many aspects of the host–parasitoid or prey–predator interaction, as we shall see in the ensuing chapters.

This book deals primarily with just one kind of natural enemy, the insect parasitoids, and attempts to review how our understanding of the dynamics of host–parasitoid interactions has advanced over the past 20 years or so. Parasitoids are amongst the most abundant of all animals, comprising some 10% or more of all metazoan species. They occur in several different insect groups, particularly in the Diptera (two-winged flies) and the Hymenoptera (sawflies, bees, wasps and ants). Within the Diptera, most species occur in two families, the Tachinidae and the Bombylidae. Within the Hymenoptera, there are about

45 families containing parasitoids including such familiar groups as the icheu-monids, braconids and chalcidoids (e.g. pteromalids and trichogrammatids). Excellent introductions to the biology of parasitoids can be found in Clausen (1940), Askew (1971) and Godfray (1994). Adult female parasitoids lay one or more of their eggs on, in or close to the body of another arthropod, usually an immature stage of another insect, which is then consumed over a period of days or weeks by the feeding parasitoid larva or larvae. As in most true parasites, all the food necessary to complete larval development comes from a single host, but like true predators this almost always leads to the death of the host, albeit with a delay until the parasitoid larva is fully developed. Parasitoid life histories can be classified by the host stage attacked; thus one can have *egg parasitoids*, *larval parasitoids* and *pupal parasitoids*, but only rarely are adult hosts attacked. Some parasitoids with arrested development attack one stage but emerge from and kill another stage; these are referred to by terms such as *egg–larval parasitoids*. Parasitoids are also classified in other ways. *Ectopara-sitoids* lay their eggs on the host's surface and the larvae then feed through small punctures in the host cuticle, while *endoparasitoids* inject their eggs directly into the host's body, and the larvae then feed internally. There are also *koinobionts* and *idiobionts* (Askew and Shaw 1986). Female koinobionts attack hosts that are too small to support the parasitoid throughout its de-velopment and the juvenile parasitoid therefore remains dormant for a period while the hosts grows. Idiobionts, on the other hand, prevent any host growth following parasitism or attack and kill a non-growing host stage. Another dis-tinction is between *solitary* parasitoids in which only one larva can complete development in a single host individual, and *gregarious* parasitoids in which a single host can support more than one parasitoid through their development. Finally, some parasitoids are known as *secondary* or *hyperparasitoids* because they attack other parasitoid species, and so differ from the usual *primary* parasitoids. Further useful terms describe parasitism of already parasitised hosts. If a previously parasitised host is attacked by a second female of the same species, *superparasitism* is said to have occurred. But, if the second female is of a different species, the term *multiparasitism* is used.

Parasitoids have been popular subjects for ecological study for a variety of reasons. First, their importance in biological pest-control programmes has stimulated a large amount of empirical and theoretical work, which seeks to identify and quantify the attributes that enhance the effectiveness of para-sitoids as pest-control agents. Second, the study of host–parasitoid population dynamics has greatly benefited from the way that parasitoids make ideal subjects for the development of relatively simple population models. This is mainly because there is only one stage, the adult females, that searches for hosts, and because the act of finding a host is then normally followed by ovi-position. The success in finding and attacking hosts therefore closely defines parasitoid reproduction. Consequently, host–parasitoid models can have a much simpler structure than corresponding predator–prey models in which all predator stages may attack prey with different effectiveness, and reproduction is less closely defined by prey consumption. This close link between parasitoid searching behaviour and reproduction has also made parasitoids a popular group amongst insect behavioural ecologists and those seeking to reveal how features of individual behaviour can affect population dynamics (Godfray 1994). Finally, many species of parasitoids and their hosts can readily be cultured in the laboratory, and this has greatly increased the amount of empirical information on parasitoids obtained under controlled conditions. There are thus several published time series for host–parasitoid interactions obtained from laboratory microcosms, and many laboratory studies exploring the ways that various factors affect parasitoid efficiency. This controlled approach to population dynamics has rightly been extolled by Kareiva (1989).

One product of the long tradition of studying the population dynamics of insects has been the number of long-term field studies that have accrued, most of which show conspicuous fluctuations in population size, but without any long-term trends in average abundances (e.g. Andrewartha and Birch 1954; Connell and Sousa 1983; Hanski 1990). The characteristic patterns of these time series fall mainly into three categories:

(1) populations showing erratic fluctuations without any trend towards increasing or declining numbers (Fig. 1.1(a));
(2) populations maintained at relatively low levels, but punctuated by occasional, spectacular episodic outbreaks, soon followed by a rapid 'crash' to former low levels (Fig. 1.1(b)); and
(3) populations showing pronounced cycles in abundance (Fig. 1.1(c)).

The mechanisms for persistence in such examples, however detailed the study, have tended to remain elusive. In particular, there has often been no clear indication that density dependence is operating, whether caused by natural enemies or other factors (Dempster 1983; Strong 1986; Gaston and Lawton 1987; Den Boer 1991), and ecologists have long disagreed over the implications of this. People have argued over definitions, over whether or not density dependence is needed for population persistence, over the frequency with which density dependence occurs in natural systems and over the best ways to detect density dependence and identify regulated populations. Despite the simplicity of the ideas, ecologists have continued 'to muddle the basic issues of the very existence of their study objects' (Hanski *et al.* 1993). The discord reached its peak in the 1950s (the Cold Spring Harbor Symposium volume of 1957 gives an excellent summary of the polarised views of the time). A. J. Nicholson (1957) and his followers maintained that populations could only persist if they were regulated by density-dependent processes, while H. G. Andrewartha (1957) and his followers denied this central role for density dependence and assumed that most populations persist by the action of various density-independent processes.

While the controversy over population regulation has been particularly

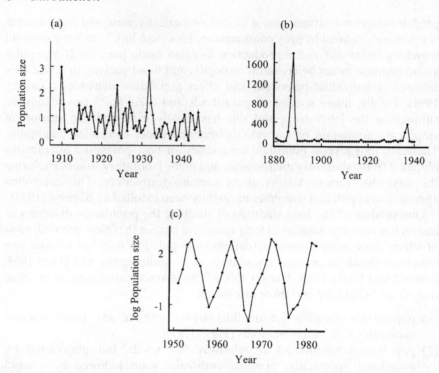

Fig. 1.1 Population fluctuations of three species of insects showing quiet different long-term dynamics: (a) the moth, *Chilo suppressalis*, caught in light traps in Okayama, Japan (data in Miyashita 1955; after Ito 1980); (b) the large pine moth, *Dendrolimus pini*, in central Germany (Schwerdtfeger 1935); (c) the larch bud moth, *Zeiraphera diniana*, in the Upper Engedine, Switzerland (Baltensweiler 1984).

intense among insect ecologists, density dependence is so obvious in some other taxa that it excites little comment. Many plants and sessile animals, for example, clearly compete for space at high population densities (Harper 1977; Tilman 1982; Paine 1984), as do territorial birds (Wiens 1989*a*); likewise, herbivorous mammals often exhaust their food supply when numbers are high, and then suffer unambiguous density-dependent mortality (Clutton Brock *et al.* 1982; Sinclair 1989). Nowadays, whatever the taxon, most ecologists would argue that any plant or animal population must experience some form of density-dependent feedback on net population growth if it is to persist for many generations in a given environment. And there is a broad consensus on the definition that a regulated population is one with a long-term, *stationary*, probability distribution of population densities, implying some mean level around which the population fluctuates (Turchin 1995). There is also now far more empirical information available, either where the actual density-dependent process has been identified and quantified or where one can infer density dependence from time series of population abundance (Turchin and

Taylor 1992; Woiwod and Hanski 1992; Hanski and Woiwod 1993; Ellner and Turchin 1995; Leirs *et al.* 1997).

This book follows on from an earlier monograph (Hassell 1978) on the components of predation and parasitism by arthropod natural enemies. These are reviewed here in Chapters 2 and 3. Chapter 2 deals with some familiar basic models for host–parasitoid interactions, and Chapter 3 refines these somewhat by including such factors as host density dependence, generalist natural enemies and density-dependent parasitoid sex ratios. Since 1978 there have been major developments in three directions: exploring the effects of age-structure on insect (including host–parasitoid) interactions formulated in continuous time, and therefore with overlapping generations (Chapter 5); examining how parasitoids may affect the dynamics of multispecies systems (Chapter 6); and determining how spatial patchiness and other forms of heterogeneity at a wide range of scales may affect population dynamics (Chapters 5 and 7).

The realisation of how important spatial processes are to population dynamics in general has led to a revolution in the subject, most of which has occurred over the past 20 years or so (Wiens 1989*b*). As a result, many of the earlier, simple host–parasitoid models, usually based on homogeneous mixing, are now viewed mainly as rather special limiting cases. As is usually the case, developing the mathematical models has generally proved less of a challenge than designing and executing appropriate studies to collect the data. It is encouraging that there are now so many examples where empirical field studies and models have been drawn close together; a few of these are reviewed later in this book (e.g. Hassell 1980a; Jones *et al.* 1993; Reeve *et al.* 1994a,b; Murdoch *et al.* 1987). Problems of spatial scale pervade ecology (Levin 1992, 1994). The interactions covered in this book are discussed at two very different spatial scales. On the one hand, there are *local populations* characterised by more-or-less complete mixing of individuals at some point(s) during the generation period. On the other hand, there are *metapopulations* formed by collections of local populations linked by some degree of dispersal each generation between the individual local populations. Most of this book is concerned with the dynamics of local populations, but these are extended to meta-populations in Chapter 7 (see Hanski 1999 for a comprehensive review of metapopulation ecology).

Much of the work on the dynamics of insect populations has taken a mechanistic approach. This is particularly true in the study of host–parasitoid interactions, and it is the approach taken in this book. Components of the interaction have been investigated one at a time, often using simple experiments, and their dynamical effects examined in a stepwise way within population models. The overall objective is to have a detailed understanding of how the important processes operating in the host and parasitoid life cycles affect the dynamics of the populations. These components may be the fundamental demographic parameters (e.g. rates of increase, searching efficiency), features of the life histories (e.g. duration and relative timing of stages susceptible to parasitism), effects of other interacting species (e.g. competing host or para-sitoid species or the effects of hyperparasitoids) or other features of the habitat (e.g. patchiness of resources, variability of resource quality). In each case empirical information is needed to describe the components and their relationship with key variables such as population density. This then yields a description of each component in an appropriate model framework. Finally, analysis of the parameterised model reveals the dynamical effects of that com-ponent. Such a mechanistic approach to population dynamics is demanding of data and requires a particularly close interaction between data and model development.

2

A simple framework

2.1 Introduction

Howard and Fiske's (1911) view that parasitoids are important in regulating host populations was probably based on their presumption of density dependence, their observations that parasitoids can cause very heavy host mortalities and their experience of some dramatically successful biological pest-control programmes, particularly the control of the cottony cushion scale (*Icerya purchasi*) by the specialist coccinellid beetle, *Rhodolia cardinalis*[1] (Koebele 1870). Since then, the potential importance of parasitoids and predators has been emphasised in a number of ways:

1. Many theoretical studies have shown how natural enemies can, in principle, regulate their host or prey populations (and much of this book is concerned with this).
2. Biological control has 'notched up' many more spectacular successes (Clausen 1940; Greathead and Greathead 1992).
3. There are a number of field studies that point to natural enemies being important factors driving the dynamics of hosts and prey (e.g. Morris 1959; Ives 1976; Munster-Swendsen 1985; Sih *et al.* 1985; Montgomery and Wallner 1988; Murdoch *et al.* 1989; Gould *et al.* 1990; Turchin 1990; Liljesthrom and Bernstein 1992; Berryman 1996).
4. Some natural enemies, particularly parasitoids, in laboratory microcosms (see, for example, Fig. 2.1) have been demonstrated unequivocally to maintain their host populations at levels well below those at which resource limitation would be important (e.g. Utida 1957; Hassell and Huffaker 1969; White and Huffaker 1969; Bellows and Hassell 1988; Hassell and May 1988; Tuda and Shimada 1995; Bonsall and Hassell 1997; Shimada 1999).

But all this work goes no further than showing that parasitoids have the potential to regulate their host populations: whether or not they are a widespread cause of the stability and persistence of natural host populations in the field is still legitimately debated. Some believe that 'top-down' regulation (host populations regulated by the stabilising effect of their natural enemies) is common in natural systems, and that the great majority of insect herbivores are too rare to be limited by their resources (e.g. Hairston *et al.* 1960; Lawton and Hassell 1984; Hawkins 1992; Godfray 1994). Others find in favour of

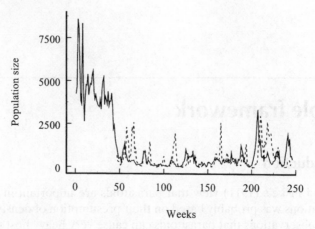

Fig. 2.1 Time series from a laboratory microcosm showing the effects on a host population of the introduction of parasitoids. The adult fruit flies, *Drosophila subobscura*
(solid line), were fed on grapes and the braconid parasitoid, *Asobara tabida* (broken
line), was introduced in week 40 of the experiment (Hassell, Reader, Bonsall and
Godfray, unpublished data).

'bottom-up' regulation (or 'donor control'), whereby host–parasitoid interactions persist primarily due to the density-dependent resource limitation of
their hosts (e.g. Dempster 1983; Hawkins *et al*. 1993; Price *et al*. 1995; Roininen
et al. 1996). Quality resources can still be limiting even if the earth appears
green (Hairston *et al*. 1960; Murdoch 1966).

This debate (revisited in Chapter 3) continues principally because of the
difficulty in getting sufficient direct and unambiguous evidence from the field.
Dempster (1983), for example, examined 24 life tables of lepidopteran species
and in only three cases were natural enemies even detected as density-
dependent factors, leaving aside the question of whether or not the density
dependence was actually the principal cause of population regulation. In a
more broad-ranging study involving 171 instances in the literature where
parasitism had been recorded from more than three host densities, Stiling
(1987) found density dependence recorded in only 25% of the cases. Such
surveys, however, are hard to interpret. The species that people have chosen
for study over the years are certainly not a representative cross-section of
host–parasitoid interactions in general. There are also several ways in which
parasitoids could be regulating their host populations and yet not be easily
detected as direct density-dependent factors. For instance, the tests used to
detect density dependence are fallible (e.g. Gaston and Lawton 1987; Holyoak
1993; Wolda and Dennis 1993; Dennis and Taper 1994; Wolda *et al*. 1994), time
delays in the density dependence can thwart some of the simple tests based on
linear models (Turchin 1990; Holyoak 1994*a,b*), and the underlying density
dependence may be obscured by stochastic effects (Hassell 1986, 1987).

Obviously, neither the 'top-down' nor 'bottom-up' model is invariably correct and the issue is one of degree rather than of kind (Hunter and Price 1992; Harrison and Cappuccino 1995; Godfray and Müller 1998). Only with fully analysed field studies, preferably accompanied by parameterised population models, will it be possible to produce a 'taxonomy' of insect systems in which either top-down or bottom-up processes predominate (Hassell *et al.* 1998).

The example shown in Fig. 2.1 emphasises two of the most basic questions about the dynamics of natural enemy–prey interactions:

1. What determines the degree to which a natural enemy depresses its average host or prey abundance below its carrying capacity?
2. By what mechanism(s) is the host or prey population maintained at these reduced, and sometimes very low, levels?

These questions have provided much of the motivation for studying host–parasitoid dynamics, and underpin much of what follows in this and later chapters.

2.2 A classical framework

Traditionally, there have been two distinct starting points for exploring the dynamics of host–parasitoid or predator–prey models, each with its own adherents. Lotka–Volterra models (Lotka 1925; Volterra 1926) start with the assumption that the generations of the interacting populations overlap completely and that birth and death processes are continuous. A separate tradition (and the one followed here) was initiated mainly by entomologists with insect hosts and their parasitoids in mind (Thompson 1924; Nicholson 1933; Varley 1947), in which the populations have discrete and synchronised generations. This model structure inevitably introduces a one-generation time lag between the act of parasitism and the resulting change in host populations, and it is the presence of these time lags that represent the fundamental difference between the two kinds of model. Such discrete-generation life cycles are widespread in temperate regions where diapause during the winter months is common.[2] Although the discrete and continuous frameworks reflect fundamentally different kinds of life cycle, both classes of model have been used to demonstrate how particular features of host–parasitoid interactions influence population dynamics.

The usual framework for discrete-generation, host–parasitoid models is given by:

$$N_{t+1} = \lambda N_t f(N_t, P_t)$$
$$P_{t+1} = cN_t [1 - f(N_t, P_t)]$$

(2.1)

where P and N are the population sizes of the searching adult female parasitoids and the susceptible host stage, respectively, in generations t and $t + 1$.

In the host equation, the parameter λ is the *net* finite rate of increase of hosts in the absence of the parasitoids. It is often assumed to be a constant, as in this chapter, both for simplicity and so that the effects of parasitism alone can be more easily discerned. Density-dependent host rates of increase are discussed in Chapter 3. The unspecified function $f(N_t, P_t)$ defines the fraction of the N_t hosts escaping parasitism; one minus this term (within the square brackets in the parasitoid equation) therefore gives the fraction of hosts parasitised. All assumptions about the efficiency of parasitoids at finding and parasitising hosts are thus contained within this term. Finally, c is the average number of adult female parasitoids emerging from each host parasitised (often assumed to be one which corresponds to parasitoids with solitary larvae). Clearly, these simple equations subsume a huge amount of host and parasitoid biology. The host rate of increase, λ, depends on the host's fecundity, sex ratio, any immigration and emigration and all host mortalities other than parasitism itself. It may also be density-dependent. The function $f(\cdot)$, and therefore the parasitoids' rate of increase, depends upon all factors affecting the level of parasitism caused by the P_t parasitoids. Lastly, the parameter c depends upon the sex ratio of the parasitoid progeny, any mortality suffered within hosts and any mortalities of the subsequent adult female parasitoids prior to searching for hosts in the next generation. The apparent simplicity of model (2.1) is thus deceptive: to be properly parameterised for a particular host–parasitoid system requires detailed life-table information on both populations (see p. 53).

The formal properties of model (2.1) are well known (e.g. Hassell and May 1973; Hassell and Pacala 1990). The populations will be at equilibrium when $P^* = cN^*(1 - 1/\lambda)$ and $1/\lambda = f(N^*, P^*)$.[3] Host and parasitoid equilibrium abundances are thus—as intuition suggests—reduced by lower host survival from parasitism, $f(N^*, P^*)$, lower host rates of increase, λ, and increased survival of parasitoid progeny, c. The equilibria are locally stable if:

$$-\frac{\lambda^2}{\lambda - 1} P^* \frac{\partial f(N^*, P^*)}{\partial P_t} < 1, \text{ and} \tag{2.2}$$

$$\frac{\lambda - 1}{\lambda} - \frac{\partial f(N^*, P^*)}{\partial P_t} > \frac{\partial f(N^*, P^*)}{\partial N_t} \tag{2.3}$$

where the derivatives define the slope of the relationship of the function f in relation to P or N close to equilibrium. Equation (2.3) will always be satisfied (because $\partial f/\partial P$ is always negative), unless in some way an increase in host abundance leads to increased host survival from parasitism. This can occur if the parasitoids are strongly egg- or handling time-limited (see Section 2.4.1 below), but is otherwise unlikely in the majority of biologically plausible models. Thus condition (2.2), which involves some kind of density dependence acting on the parasitoid population (e.g. their per capita efficiency declines as parasitoid density increases—see Section 2.4.2), is usually the necessary and sufficient condition for stability.

Model (2.1) has been used to explore the ways that many different components of a host–parasitoid interaction can affect dynamics, including:

(1) host density (Section 2.4.1);
(2) parasitoid density (Section 2.4.2);
(3) non-random search (Section 2.3.3 and Chapter 4);
(4) shifting sex ratios in the parasitoid population (Chapter 3);
(5) density-dependent parasitoid survival (Chapter 3);
(6) density-dependent rates of increase for the host population (Chapter 3);
(7) gregarious, rather than solitary, parasitoid larvae (Chapter 3);
(8) explicit age-structure for the populations (Chapter 5);
(9) additional host and natural enemy species (Chapter 6); and
(10) spatially explicit interactions (Chapters 5 and 7).

But first, we turn to some basic models that set much of this work in train.

2.3 Three models

2.3.1 W. R. Thompson

As an entomologist interested in the release of parasitoids for the biological control of insect pests, W. R. Thompson was particularly concerned with the performance of relatively few parasitoids faced with a 'sea' of hosts. This perhaps explains his assumption that female parasitoids are not limited by their ability to find hosts, but rather by their egg supply. Each female would then lay a constant number of eggs, w, reflecting her effective egg complement (Thompson 1924; Thompson 1930). In his simplest models, these w eggs were invariably laid in w different hosts. But Thompson realised that parasitoids are unlikely to be as efficient as this, and in other models he included random superparasitism[4] of hosts. Each parasitoid still laid w eggs, but these were 'scattered' randomly amongst the available hosts, following a Poisson distribution. The probability of a host escaping parasitism, $f(\cdot)$ in model (2.1), is thus given by the zero term of the Poisson as shown in Row A of Table 2.1.

The dynamics of this model are difficult to relate to the real world. Depending on the values of w, λ (assumed constant) and the initial host and parasitoid densities, the host population either 'escapes' from the parasitoids with both populations increasing unchecked, or the parasitoid drives the hosts to extinction.[5] Only if additional density dependence acting on the hosts or parasitoids is introduced is a stable host–parasitoid equilibrium possible.

Thompson's model is thus one of random exploitation of hosts by parasitoids, in which finding enough hosts is not an issue—the available eggs are always laid, irrespective of host density. It may apply quite well, therefore, during the early stages of a biological-control programme when introduced parasitoids are faced with superabundant hosts. We would also expect individual parasitoids sometimes to run out of available eggs. For example,

Table 2.1 Some expressions for natural enemy functional responses from the literature. In column 3 host individuals are not removed after attack so that re-encounters can occur. N_{enc} therefore gives the total number of encounters with N hosts by P searching parasitoids. Column 4, however, gives the fraction of hosts parasitised, $f(\cdot)$, irrespective of the number of times they are encountered. A distribution of attacks therefore needs to be specified, which in the examples below is either the Poisson or negative binomial distributions. (Note that, for convenience, the generation subscript t is omitted and the total searching time, $T = 1$.)

Row	Comments	Hosts encountered by P parasitoids, N_{enc}	Fraction of hosts surviving $f(\cdot)$	Author(s)
A	Constant number of hosts encountered per parasitoid; random distribution of attacks	wP	$1 - \exp\left(-w\dfrac{P}{N}\right)$	Thompson (1924)
B	Linear response; random attacks	aNP	$1 - \exp(-aP)$	Nicholson (1933); Nicholson and Bailey (1935)
C	Linear response; negative binomial distribution of attacks	aNP	$1 - \left(1 + \dfrac{aP}{k}\right)^{-k}$	Griffiths (1969b); May (1978)
D	Type II parasitoids with random attacks	$\dfrac{aNP}{1 + aT_hN}$	$1 - \exp\left(-\dfrac{aP}{1 + aT_hN}\right)$	Holling (1959a,b); Royama (1971); Rogers (1972)
E	Type II predators with random attacks (N_a is the total number of prey attacked)	as in D	$1 - \exp[-a(P - T_hN_a)]$	Royama (1971); Rogers (1972)
F	Type II parasitoids with negative binomial distribution of attacks	as in D	$1 - \left(1 + \dfrac{aP}{k(1 + aT_hN)}\right)^{-k}$	Hassell (1980a)
G	Type II response with separate handling time (T_p) for already-parasitised hosts. Row D recovered if $T_p = T_h$; Row E recovered if $T_p = 0$	$\dfrac{a[P - (T_h - T_p)N_a]}{1 + aT_hN}$	$1 - \exp\left[-\dfrac{a(P - (T_h - T_p)N_a)}{1 + aT_hN}\right]$	Arditi (1983)

H	Type III response where $a = bN/(1+cN)$	$\dfrac{bN^2P}{1+cN+bT_hN^2}$	$1-\exp\left(-\dfrac{bNP}{1+cN+bT_hN^2}\right)$	Hassell et al. (1976)
I	As H, but with separate handling time (T_p) for already parasitised hosts. Row H (parasitoids) recovered if $T_p = T_h$; predator response obtained if $T_p = 0$	$\dfrac{bN[P-(T_h-T_p)N_a]}{1+cN+bT_pN^2}$	$1-\exp\left(-\dfrac{bN[P-(T_h-T_p)N_a]}{1+cN+bT_pN^2}\right)$	Arditi (1983)
J	Response with flexibility between the limits of being search-limited to egg-limited; converges on row B (search-limited) when β $(=1/T_h) \to \infty$ and on row A (egg-limited) when $a \to \infty$	$\dfrac{a\beta NP}{\beta+aN}$	$1-\left(1+\dfrac{a\beta P}{k(\beta+aN)}\right)^{-k}$	Getz and Mills (1997)
K	Combining Type II functional response with interference	$\dfrac{aNP^{-m}}{1+aT_hNP^{-m}}$	$1-\exp\left(-\dfrac{aP^{-m}}{1+aT_hNP^{-m}}\right)$	Arditi and Akçakaya (1990)

Driessen and Hemerik (1992) predicted that some 13% of females of the pro-ovigenic[6] eucoilid parasitoid, *Leptopilina clavipes*, attacking *Drosophila* larvae exhausted their supply of eggs before dying. Synovigenic parasitoids are also likely to be egg-limited on occasion when hosts are plentiful. In general, however, most workers in the field have assumed that parasitoid populations, as a whole, are more likely to be search-limited than egg-limited, and that host density will thus strongly influence the number of eggs that a female parasitoid will lay.

2.3.2 The Nicholson–Bailey model

Another entomologist, A. J. Nicholson, has had a much greater impact on the development of population ecology than Thompson—for his championing of the concept of population regulation by density-dependent factors (Nicholson 1947), for his models of host–parasitoid dynamics (Nicholson 1933; Nicholson and Bailey 1935) and for distinguishing between scramble and contest competition (Nicholson 1954). His career and many contributions have been reviewed by Hopper (1990), Mackerras (1970) and Kingsland (1996). Together with V. A. Bailey (a physicist), Nicholson explored a version of model (2.1) in depth, in which the following assumptions about parasitism were made.

1. The parasitoids are never egg-limited and encounter hosts in direct proportion to host abundance. The total number of encounters with hosts is therefore given by $N_{enc} = aN_tP_t$ (in contrast to $N_{enc} = wP_t$ from Thompson's model).[7] The constant a is the per capita searching efficiency which Nicholson called the 'area of discovery', because he assumed that individual parasitoids of a particular species typically cover a characteristic area in their searching lifetimes within which all hosts would be parasitised, however abundant they may be.

2. As assumed by Thompson, the N_{enc} encounters are distributed randomly amongst the hosts, all of which are equally susceptible. Thus, there is no avoidance of superparasitism if an egg is laid at each encounter. Alternatively, if the parasitoids *can* avoid superparasitism they do so instantaneously without affecting subsequent performance in any way; for example by 'wasting time' handling already parasitised hosts.

These assumptions differ little from those in the Lotka–Volterra predator–prey model; the primary difference is that they are now embedded in an interaction with discrete rather than continuous generations. Graphically, Nicholson represented his assumptions about parasitism in his so-called 'competition curve' (Fig. 2.2) in which the proportion of hosts escaping parasitism is given by the zero term of the Poisson distribution, $\exp(-aP_t)$, where aP_t are the mean encounters per host, $N_{enc}/N_t = aP_t$. Thus one minus this zero term is the probability of a host being attacked.

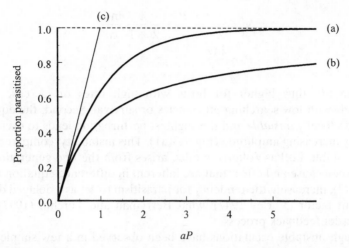

Fig. 2.2 Nicholson's 'competition curve', line (a), obtained from eqn (2.5) in which a is the parasitoid searching efficiency ('area of discovery') and P is the density of searching adult parasitoids. The line shows the proportion of hosts parasitised rising monotonically towards 100% parasitism. This is compared with line (b) in which parasitism is obtained from the negative binomial model (2.7) with $k = 0.7$, and with the straight line (c) which would arise if the parasitoids 'co-operated' in some way to avoid encountering any host more than once.

Substituting into model (2.1) gives:

$$N_{t+1} = \lambda N_t \exp(-aP_t)$$
$$P_{t+1} = N_a = cN_t[1 - \exp(-aP_t)] \tag{2.4}$$

where N_a is the number of hosts which are parasitised irrespective of the number of times they have been encountered. By rearranging the parasitoid equation, and assuming $c = 1$, we can see that the per capita searching efficiency of the parasitoids, which is the proportion of total hosts encountered by parasitoids per unit time, can be represented by[8]:

$$a = \frac{1}{P_t} \ln\left[\frac{N_t}{N_t - N_a}\right]. \tag{2.5}$$

This is important because it shows that one can readily estimate searching efficiency (assuming random exploitation of hosts) whenever data on the proportion of hosts parasitised by a known number of parasitoids is at hand.[9]

The dynamical properties of the Nicholson–Bailey model can be readily portrayed (Hassell and May 1973). First, unlike the Thompson model, a host–parasitoid equilibrium always exists depending on the values of a and λ. By setting $N_{t+1} = N_t = N^*$ and $P_{t+1} = P_t = P^*$, we have:

$$N^* = \frac{\lambda}{\lambda - 1} \frac{P^*}{c} = \frac{\lambda}{\lambda - 1} \frac{\ln \lambda}{ca}$$

$$P^* = \frac{\ln \lambda}{a}. \tag{2.6}$$

Equilibria are thus higher for hosts with high rates of increase and/or parasitoids with low searching efficiencies or survival. Second, the equilibria are always locally *unstable* and the slightest perturbation leads to oscillations of rapidly increasing amplitude (Fig. 2.3(a)). This instability, compared to the neutrally stable Lotka–Volterra model, arises from the one-generation time lags between cause and effect that are inherent in difference equation models (May 1973), increasing the tendency for parasitism to act as a delayed density-dependent factor (Varley 1947): what Berryman and Turchin (1997) call a second-order feedback process.

Although unstable oscillations have been observed in a few simple laboratory host–parasitoid and predator–prey experiments (e.g. Figs 2.3(b), (c)), such instability is hard to reconcile with the results from other laboratory systems in which the interactions are much more stable (e.g. Utida 1957; Huffaker 1958; Huffaker *et al.* 1963; Fujii 1983; Bonsall and Hassell 1997; 1998) (and see Fig. 2.1) and, more generally, with the long-term persistence of natural systems. Nicholson (1947), anticipating the current vogue for meta-populations, suggested one means of persistence of oscillatorily unstable local populations. Assuming that the interaction occurs in distinct and separated areas, the 'cycle of increase in numbers, followed by ... extermination, proceeds independently in different parts of the occupied country; so at all times some groups are increasing and some decreasing in numbers ... Consequently when one considers a large tract of country, the abundance [of both host and parasitoid] ... remains more or less constant; whereas in any small area of the same country the fluctuation in numbers ... may be violent.' Such meta-population persistence is considered in detail in Chapter 7.

There are, however, several other ways in which the Nicholson–Bailey model can be modified that both add realism and allow the populations to persist. These involve elaborating the parasitism function $f(\cdot)$ (see Section 2.4 and Chapter 4), allowing the host rate of increase, parasitoid survival or parasitoid sex ratios to be density-dependent (Chapter 3), or adopting more realistic life histories for both hosts and parasitoids (Chapter 5).

2.3.3 The negative binomial model

The assumption that hosts are parasitised at random implies equal susceptibility to parasitism amongst all host individuals—an unlikely proposition in the real world where host individuals are bound to vary in their spatial location, phenotype and stage of development. Much more likely, therefore, is that the risk of being parasitised will vary within the host population, leading

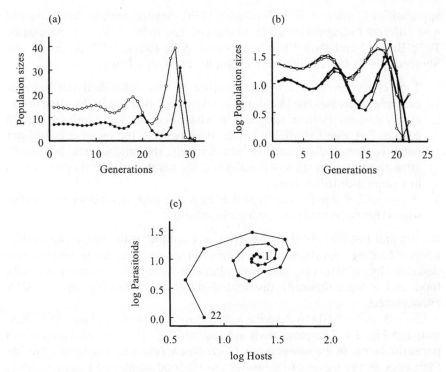

Fig. 2.3 (a) Numerical simulation showing host (hollow circles) and parasitoid (solid circles) population oscillations from the Nicholson–Bailey model (2.4) where the parasitoid searching efficiency $a = 0.02$ and the host rate of increase $\lambda = 2$. (b) Population fluctuations from a laboratory interaction between the greenhouse whitefly, *Trialeurodes vaporariorum* (hollow circles), and its chalcidid parasitoid, *Encarsia formosa*, (solid circles). The heavy lines show the observed populations and the fine lines show the Nicholson–Bailey predictions where $a = 0.068$ (the mean value over the 22 generations) and $\lambda = 2$ (the value imposed by the experimental design) (Burnett 1958). (c) Phase space plot of the observed populations from (b) emphasising the delayed density-dependent parasitism. The first and last generations are marked.

to an overall distribution of parasitoid attacks that is more aggregated than random (this is discussed more fully in Chapter 4). Amongst the wide range of statistical descriptions for such clumped events, the negative binomial distribution (NBD) is particularly popular with ecologists. It is defined by two parameters: the mean of the distribution and a parameter, k, that inversely defines the extent of clumping (most aggregated as $k \to 0$, becoming random (i.e. Poisson) as $k \to \infty$). It is thus a flexible distribution and describes well the degree of clumping in the spatial distribution of a wide range of natural populations (e.g. Lyons 1964; Ibarra *et al.* 1965; Anderson and May 1978; Atkinson and Shorrocks 1984; Southwood *et al.* 1989; Hails and Crawley 1992; Naeem and Fenchel 1994), although there are problems in some of its applications (Taylor *et al.* 1979; Green 1986). Several authors have stressed how different biological mechanisms can give rise to the NBD (e.g. Anscombe 1959; Boswell and Patil 1970; Southwood 1976; Elliott 1977; Atkinson and Shorrocks 1984; Green 1986). For example, it can arise from:

1. *True contagion* in which the presence of one individual (an egg, for example) increases the likelihood of finding another egg in the same place.
2. *Heterogeneous Poisson sampling* in which eggs are laid randomly within patches but, due to variable patch attractiveness, the mean eggs laid per patch is gamma-distributed or, alternatively, the patches may be equally attractive but the quality of females (i.e. the number of eggs they lay) varies in a gamma-distributed way.
3. *A compound distribution* in which eggs are laid at random in clutches which themselves vary in size logarithmically.

Naeem and Fenchel (1994) have given an example of the latter: the distribution of feeding opportunities by a marine ciliate on wounded invertebrates is described by an NBD obtained from a Poisson distribution of encounters with food and a logarithmically distributed duration of feeding once food is encountered.

The use of the NBD to describe a distribution of parasitoid attacks is illustrated in Fig. 2.4. The parasitoids are tachinid flies (*Cyzenis albicans*) which parasitise larvae of the winter moth, *Operophtera brumata*. The female flies lay their eggs on the leaves of the winter moth's food plant, and parasitism then occurs when an egg is ingested by a feeding winter moth caterpillar. The eggs

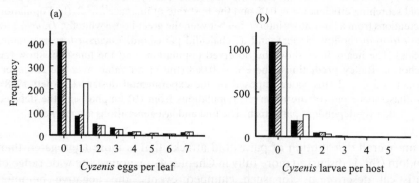

Fig. 2.4 (a) Frequency distribution of eggs laid per leaf by the tachinid fly, *Cyzenis albicans*. Black columns show the observed data and hatched columns the expected frequencies from a negative binomial distribution with best fit estimate of the clumping parameter $k = 0.28$. The hollow columns are the expected frequencies from the Poisson distribution and clearly provide a much poorer fit to the data. (b) As for (a), but now showing the frequencies of winter moth (*Operophtera brumata*) larvae containing different numbers of first instar *Cyzenis* larvae. The expected frequencies now come from the negative binomial distribution with estimated $k = 0.55$. Hollow columns are again the expected frequencies from the Poisson distribution (Hassell and May 1985).

are not, however, just scattered randomly; the parasitoids respond to the sap fluxes from damaged leaves by ovipositing nearby (Hassell 1968). Winter moths at higher densities therefore attract more parasitoids and suffer higher levels of parasitism than winter moths at lower densities. For other examples, see Driessen and Hemerik (1991) and Reeve, Cronin and Strong (1994).

With such examples in mind, May (1978), modified the Nicholson–Bailey model by assuming that host survival from parasitism is given by the zero term of the NBD (Row C; Table 2.1) instead of the zero term of the Poisson distribution (see also Griffiths (1969b) and Griffiths and Holling (1969)):

$$N_{t+1} = \lambda N_t \left[1 + \frac{aP_t}{k}\right]^{-k}$$

$$P_{t+1} = cN_t \left(1 - \left[1 + \frac{aP_t}{k}\right]^{-k}\right).$$

(2.7)

The full derivation of this model is given in Chapter 4, but note that small values of k indicate strong clumping in the distribution of parasitoid attacks, and that as $k \to \infty$ the Poisson distribution, and hence the Nicholson–Bailey model, is recovered.[10]

The properties of model (2.7) are importantly different from those of the Nicholson–Bailey model. First, the equilibrium populations are given by:

$$N^* = \frac{\lambda P^*}{c(\lambda - 1)}$$

$$P^* = \left(\frac{k}{a}\right)\left(\lambda^{\frac{1}{k}} - 1\right)$$

(2.8)

and thus do not just depend on λ, c and a, but also on k. For very small values of k, the model predicts high host equilibrium densities, at which the host population could well be resource-, rather than parasitoid-limited, so that λ should be density-dependent rather than constant. (The interplay between parasitism and host density dependence is further explored in Chapter 3.) Second, the equilibria are locally stable provided that the distribution of parasitism is sufficiently aggregated, or, more specifically, if, and only if, $k < 1$ (May 1978). A numerical example is given in Fig. 2.5. Values of k larger than one lead to expanding oscillations. The stabilising effect of small k stems from the way that the aggregation of attacks amongst the host population leads to a density-dependent relationship between the density of adult parasitoids per generation and their per capita searching efficiency, as illustrated in Fig. 2.2 (see also p. 29).

As Taylor (1993a) has emphasised, simple stability criteria can sometimes hide a wealth of interesting dynamics. For example, the stability criterion $k < 1$ is given by the condition for the absolute value of the dominant eigenvalue, $|\psi|$, being less than one. However, ψ also measures other aspects of stability,

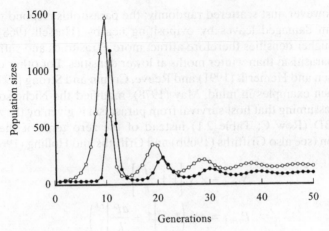

Fig. 2.5 Numerical example from the negative binomial model (2.7) of May (1978). The parameters are the same as in Fig. 2.3(a) (i.e. $a = 0.068$ and $\lambda = 2$), except for parasitism no longer being random ($k = 0.6$ instead of $k \to \infty$).

and examining the precise relationship between $|\psi|$ and k reveals further interesting properties within the stable region (Fig. 2.6). For example, one aspect of stability, the rate at which the populations return to the equilibrium, depends upon the 'characteristic return time' (Beddington *et al.* 1976), defined in this case by $T_R = (1/(1 - |\psi|))$. The value of T_R is increased as k decreases, but only up to a point, and to the left of the cusp in Fig. 2.6 it declines as $|\psi| \to 1$. Also, whether ψ is complex or real determines the kind of approach of the populations to equilibrium: as k decreases, the populations show oscillatory damping to the right of the cusp (complex ψ) but monotonic damping to the left of the cusp (real ψ). In short, damped oscillations are associated with a more rapid return to equilibrium than monotonic damping.

2.4 Functional responses and interference

In the almost complete absence of detailed empirical information, early models of host–parasitoid interactions described the outcome of parasitoid searching behaviour in terms of a single constant, the average fecundity per female parasitoid, w, (Thompson 1924) or the parasitoid searching efficiency, a, (Nicholson 1933). Not surprisingly, as data has accumulated, it has become increasingly obvious that there are other important influences determining levels of parasitism, and that the per capita efficiency of parasitism is influenced in systematic ways by host and parasitoid abundance that were not considered by Thompson or Nicholson. Some of these responses to host and parasitoid density and their dynamical consequences are outlined here.

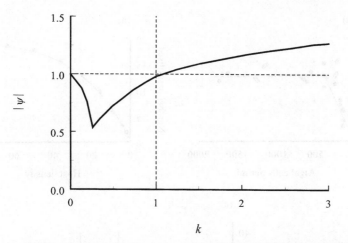

Fig. 2.6 Stability in terms of the absolute value of the dominant eigenvalue ($|\psi|$) from May's (1978) negative binomial model (2.7) in relation to the degree of aggregation of parasitoid attacks (k). The curved line shows how stability changes with k for a fixed host rate of increase, r ($= \ln\lambda$) $= 2$. In the region $k < 1$, monotonic damping occurs to the left of the minimum where ψ is real, and oscillatory dynamics to the right of the cusp where it is complex. (After Taylor 1993a.)

2.4.1 Host density—functional responses

The term 'functional response' was originally coined by Solomon (1949) for the relationship describing the way that prey density affects the number of prey attacked per predator per unit time. Functional responses are therefore central to any description of parasitism or predation. Subsequently, in two classic papers in 1959, C. S. Holling emphasised the importance of these relationships (1959*a*, *b*), and proposed a classification based on three basic types of functional responses (Fig. 2.7). The type I functional response (Fig. 2.7(a)) rises linearly with host density (with the slope equal to the searching efficiency) as in Row B of Table 2.1, but then abruptly reaches a plateau in which the maximum attack rate is constant for all further increases in host density (as in Row A). Type I responses are therefore a hybrid of the more realistic aspects of both the Nicholson–Bailey and Thompson models. Holling rightly argued that filter feeders would come closest to this ideal, provided that they filter food at a constant rate while feeding and cease to feed abruptly once satiated (e.g. Rigler 1961).

Holling's type II response (Fig. 2.7(b)), in addition to the searching constant, a, has the added ingredient of what Holling called the 'handling time', T_h, which is the time interval between first detecting a host individual and subsequently resuming the search for other hosts. Because handling time reduces the time available for searching, the resulting response rises at a decreasing rate as more hosts are parasitised, towards an upper asymptote—the maximum

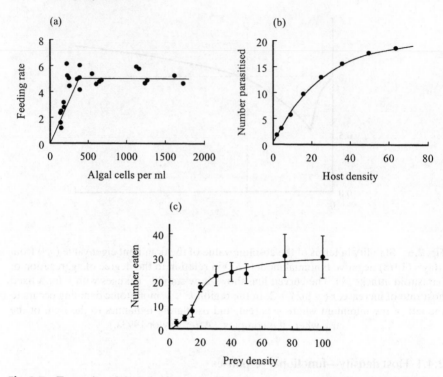

Fig. 2.7 Examples of three basic types of functional response (Holling 1959*a*); all lines fitted by eye. (a) A type I response for *Daphnia magna* feeding on different densities of algal cells. The response rises linearly to an abrupt plateau, at which point the predator is satiated and ceases feeding (Rigler 1961). (b) A type II response for the parasitoid, *Dahlbominus fuscipennis*, attacking pupae of the sawfly, *Neodiprion sertifer* (Burnett 1956). (c) Sigmoid, type III, response for the waterboatman, *Plea atomaria*, feeding on mosquito larvae, *Aedes aegypti* (A. Reeve, unpublished data). (From Hassell *et al*. 1977.)

attack rate (T/T_h), at which all the available time is spent 'handling' attacked hosts leaving no time left for searching. Long handling times therefore lead to low maximum attack rates and vice versa. The so-called 'disc equation' for type II responses (named after Holling's (1959*b*) experiment in which blind-folded subjects acted as predators searching for sandpaper discs as prey) is shown in Row D, Table 2.1. Nowadays, it is generally recognised that a constant handling time *per se* is only one mechanism producing type II responses, and the same relationship can be produced from several models (e.g. Ivlev 1961), for example, due to satiation.

Finally, Holling's type III response is sigmoid (Fig. 2.7(c)). As host density rises, the response initially accelerates due to the parasitoid or predator becoming increasingly efficient at finding hosts or prey (*a* increases and/or T_h decreases—see Row H, Table 2.1). It then levels off under the influence of handling time or satiation. Holling (1959a) assumed that type II functional responses were typical of invertebrates, while these type III responses were more typical of vertebrate predators that have greater abilities to respond to increasing prey densities by improved efficiency (e.g. learning precisely where to search to best effect). The available evidence, however, does not support this view; there are at least as many examples of sigmoid responses from insect predators and parasitoids as from vertebrates (e.g. Takahashi 1968; Murdoch and Oaten 1975; Hassell *et al.* 1976a, 1977; van Lenteren and Bakker 1976).

Holling's terminology is now almost universally adopted and adequately pigeonholes most functional responses that have been quantified. The main exceptions are some dome-shaped responses that may arise from a 'confusion' effect (Holling 1961), a 'disturbance' effect (Mori and Chant 1966) or a 'prey defence' effect (Tostawaryk 1972) at high host or prey densities. Most of this empirical information on functional responses comes from laboratory experiments, in which a fixed number of searching parasitoids or predators are confined with a particular host density for a fixed period of time. The whole response is obtained by repeating the experiment over a sufficiently broad range of host densities. An alternative design recognises the artificiality of confining a parasitoid with hosts for a fixed period, irrespective of the length of time that it would 'choose' to remain on the container of hosts were it unconstrained. To allow for this, some experiments and analyses have been of variable duration, depending on when the parasitoid leaves the host container (e.g. van Lenteren and Bakker 1976; Casas *et al.* 1993). Parasitoids usually move freely between patches in a patchy environment, making it important to consider the time that parasitoids spend on patches of different host density. This is considered in more detail in Chapter 5.

Each of the functional responses in Table 2.1 is displayed in two ways. Column 3 gives the number of encounters with hosts, N_{enc}, assuming that the host individuals are not removed after attack, so that re-encounters can occur (this is equal to the number of eggs laid if there were no avoidance of superparasitism). Column 4 shows the fraction of hosts parasitised, irrespective of the number of times they are encountered (this is equivalent to the fraction of prey eaten by predators assuming that consumed prey are not immediately replaced). The way that encounters with hosts are distributed therefore needs to be specified; for example, with a Poisson distribution if hosts are encountered at random, or with the negative binomial distribution if encountered in an aggregated way. All these expressions can readily be fitted to real functional response data using suitable non-linear least squares or maximum likelihood methods to estimate the key parameters. The relative merits of different fitting procedures have been widely discussed (e.g. Glass 1970; Rogers 1972; Arditi 1983; Juliano and Williams 1987; Trexler *et al.* 1988; Juliano 1993; Casas and Hulliger 1994; Fan and Petitt 1994; Williams and Juliano 1996).

The proliferation of functional response models is partly due to the differences between parasitoids and predators; predators consume their prey (which can thus only be encountered once), while parasitoids can re-encounter their hosts and therefore may spend more time in total 'handling' them. Originally, Royama (1971) and Rogers (1972) formulated different type II responses for each category[11] (see Rows D and E, Table 2.1). Subsequently, Arditi (1983) produced a single, more flexible, version embracing both predators and parasitoids (row G), in which an additional parameter, T_p, is introduced to represent the time parasitoids spend handling hosts that have already been parasitised (see also Arditi and Glaizot 1995). The original predator response is now recovered if $T_p = 0$; the original parasitoid response is recovered if $T_p = T_h$, with all shades allowed in-between for parasitoids that spend shorter times handling parasitised than healthy hosts. The comparable treatment for type III responses is given in Row I (Arditi 1983).

The impact that these different kinds of functional response can have on population dynamics is easily assessed by substituting the different expressions for host survival, $f(\cdot)$, from Table 2.1 into model (2.1) and comparing the results with the appropriate 'null' model—the Nicholson–Bailey model when the distribution of encounters is random, or May's model when the distribution follows the negative binomial. Type II responses are inversely density-dependent throughout their range of host densities, and are therefore increasingly destabilising as the ratio T/T_h increases (i.e. as the maximum attack rate or upper asymptote of the response is lowered). At the same time, however, the overall efficiency of the parasitoids is also reduced (compared to the linear, Nicholson–Bailey responses), thereby increasing the equilibrium levels of the populations and making it more likely that the hosts will be competing for resources and therefore adding stability to the interaction.[12] When the type II response is combined with a negative binomial distribution of host encounters, there will be an additional interplay between the stabilising effects of aggregated parasitism (small k) and the destabilising effects of the functional response *per se*. In particular, the tendency for small values of k to raise the host equilibrium (see above) has the effect of increasing the total time at equilibrium spent handling hosts, which in turn decreases stability (Hassell 1984b).

For most parasitoids handling time, in the narrow sense, is likely to make up a relatively small part of the total searching lifetime (T_h/T is small), tending therefore to make the responses more linear and the destabilising effects relatively small (Hassell and May 1973). Sometimes, however, parasitoids will run out of eggs which will be the main factor limiting their attack rates. Type II responses may therefore be more important dynamically than an analysis based on handling times alone would suggest. An interesting development along these lines is given by Getz and Mills (1997) (see row J). They formulate a functional response model which has the limits of parasitoids either being egg-limited (Thompson 1924) or being search-limited (Nicholson 1933). Assuming a negative binomial distribution of encounters, they show how the requirement for stability of $k < 1$ with a linear functional response (see p. 19) is progressively restricted as the parasitoids become increasingly egg-limited and hence the functional response becomes more acutely 'type II' in shape.

Type III functional responses (row H), in contrast, are density-dependent over the accelerating part of the curve and therefore contribute to stability, at least at relatively low host densities. This stabilising effect, however, is relatively weak for coupled host–parasitoid systems of the form of model (2.1) with its destabilising one-generation time delay (Hassell and Comins 1978). It is not possible, therefore, to stabilise the Nicholson–Bailey model (2.4) just by substituting a type III parasitoid response. Only if the time delays are reduced or absent, as in continuous-time models (Murdoch and Oaten 1975), or if the parasitoid dynamics are uncoupled from their hosts (see Chapter 3) do these sigmoid responses on their own have a marked effect in stabilising the interactions.

Table 2.1 shows a range of functional response models of varying complexity. While we can assess the impact on population dynamics of each in isolation, a better understanding of this comes when they are embedded in models in which there is a spatial distribution of host densities per patch. This is discussed in Chapter 5.

2.4.2 Parasitoid density—interference

Mutual interference

In some parasitoid species, adults react noticeably to other searching individuals nearby. Searching may be interrupted for a period, after which it may be resumed or one, or both, parasitoids may then disperse from that area of hosts (e.g. Waage 1979; Lawrence 1981; Mills 1991). Sometimes, similar behaviour is observed when an adult parasitoid detects a host that is already parasitised (e.g. Rogers and Hassell 1974; Roitberg and Prokopy 1987) or detects a chemical trail left by a parasitoid previously in that area (e.g. Price 1970, 1972; Vinson 1972). One consequence of all these behaviours is that the time actually available for search may decrease as parasitoid density increases. Such mutual interference between parasitoids has most often been observed from laboratory experiments in which parasitoid density has been varied but host density fixed. In this way, it has been easy to quantify interference in terms of the rate of decline in searching efficiency, measured from eqn (2.5), as parasitoid density increases (Fig. 2.8).

Examples such as these led Watt (1959) and Hassell and Varley (1969) to develop similar descriptive models for parasitoid interference. The Hassell–Varley model is simply a modification of the Nicholson–Bailey model based on the observation that relationships such as those in Fig. 2.8 are well described by $a = QP_t^{-m}$ where Q is the searching efficiency when $P_t = 1$ and m is the

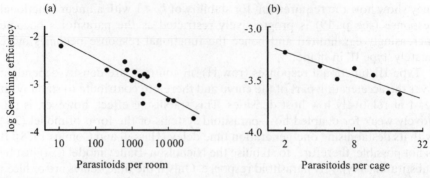

Fig. 2.8 Examples of linear mutual interference relationships between searching efficiency (log *a*) calculated from eqn (2.5) and the density of searching parasitoids. (a) The ichneumonid parasitoid *Venturia canescens* parasitising larvae of the flour moth, *Ephestia kühniella*. Data collected over 15 generations from an interaction in a laboratory controlled-environment room (Hassell and Huffaker 1969). (b) The ichneumonid parasitoid *Cryptus inornatus* parasitising cocoons of *Loxostege sticticalis* in a laboratory cage system (Ullyett 1949).

slope of the log–log relationship called the 'interference constant'. Host survival is therefore given by:

$$f(P_t) = \exp(- QP_t^{1-m}) \tag{2.9}$$

and the Nicholson–Bailey model is recovered if $m = 0$, when searching efficiency becomes once again independent of parasitoid density.[13] The stability properties of model (2.1) with *f* from eqn (2.9) are shown in Fig. 2.9. As the strength of interference, *m*, increases the density-dependent reduction in parasitoid searching efficiency (the average across all parasitoids in the population) becomes stronger and hence the stabilising effect gets greater. Thus, as long as interference is not too great ($m < 1$), there is a broad region in which the interaction is stable. However, with values of $m > 1$ there is so much parasitoid density dependence that a host–parasitoid equilibrium can no longer exist and the host population 'escapes' from parasitism. Note that this analysis assumes linear functional responses; if interference is combined with type II responses, the stable region is progressively reduced as the handling time parameter is increased (Hassell and May 1973).

While some interference relationships are approximately linear over the range of parasitoid densities used, this is certainly not true of all of them, as shown by the examples in Figs 2.10(a), (b) (see also Shimada 1999), and in any event cannot continue to be true when parasitoid densities become sufficiently small for interference to become negligible (Royama 1971). There have been a number of alternative, and more realistic, descriptions of mutual interference with a more mechanistic basis that predict such patterns (Rogers and Hassell 1974; Beddington 1975; Free *et al.* 1977). In the same vein as Holling's (1959*b*)

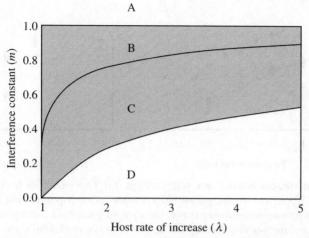

Fig. 2.9 Stability boundaries from the linear interference model (eqn (2.9)) in terms of the mutual interference constant (m) and the host rate of increase (λ). The model is locally stable within the shaded area, which is divided into two regions: B in which there is an exponential approach to equilibrium and C with oscillatory damping. Below this, in region D, the model is oscillatorily unstable (the Nicholson–Bailey model is recovered when $m = 0$), while in region A the interference is so strong that there is no equilibrium and the parasitoids are unable to prevent the continual increase of the host population. (After Hassell and May 1973.)

disc equation, Beddington (1975) defined an encounter rate b between foraging parasitoids and a parameter T_w for the proportion of total time 'wasted' from each encounter, to give:

$$f(P_t) = \exp\left(-\frac{aP_t}{1 + bT_w(P_t - 1)}\right). \tag{2.10}$$

Here $bT_w(P_t - 1)$ represents the proportion of the total time wasted due to interference, and the Nicholson–Bailey model is recovered when interference is absent ($bT_w = 0$). As interference increases, the relationships become increasingly curvilinear. The dynamical effects of these curvilinear relationships hinge on the slope, m^*, of the relationship evaluated at the *equilibrium* parasitoid density, obtained in this case from $m^* = bT_w\log_e\lambda/a$. The contribution to stability can now be evaluated by using m^* and λ in Fig. 2.9.

Mutual interference clearly has the potential to have a major impact on the stability of host–parasitoid systems. That it is important in the real world, however, is less likely. Just as the impact of handling time on stability depends on the magnitude of T_h in relation to the total time available, T, (see above), Free *et al.* (1977) have shown that the impact of interference from eqn (2.10) depends on the ratio T_w/T evaluated at the equilibrium. This ratio needs to be relatively high for interference to have an important effect on stability. For

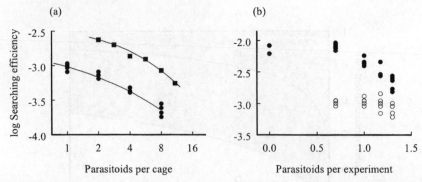

Fig. 2.10 Curvilinear interference relationships. (a) Two examples from laboratory experiments. Upper curve: the parasitoid *Encarsia formosa* parasitising the whitefly *Trialeurodes vaporariorum*; lower curve: the cynipid parasitoid *Leptopilina* (= *Pseudeucoila*) *bochei* parasitising *Drosophila* larvae (Bakker *et al.* 1967). (b) Contrasting results from experiments in the laboratory and in field cages for the cynipid parasitoid *Trybliographa rapae* parasitising larvae of the cabbage root fly, *Delia radicum*. The upper data (solid circles) come from an experiment in plastic arenas ($45 \times 45 \times 10$ cm) over a 24-hour period in which hosts were fed on discs of swede. The lower data (hollow circles) come from a field experiment in net-covered cages ($100 \times 100 \times 50$ cm) over a 48-hour period in which hosts were fed on small swede plants. (After Jones and Hassell 1988.)

example, values of $m^* > 0.5$ require that at equilibrium a single encounter with another parasitoid 'wastes' more than half of the parasitoid's potential searching time, T. In other words, and not surprisingly, the density-dependent reduction in searching efficiency will be much greater the more that parasitoids are confined within relatively small cages compared to more natural conditions (Fig. 2.10(b)). Mutual interference is more likely to be important when parasitoid populations are well above their equilibria; for example, following their successful introduction in a biological-control programme. It may then contribute to the rapid dispersal of the introduced parasitoids that is a feature of many of these programmes (Townes 1971; Hassell 1978). Interference may also be important in patchy environments where searching parasitoids tend to aggregate in some patches rather than others (Fig. 2.11). The high levels of mutual encounters in these patches, and the resulting increased dispersal, can be important to the way that the parasitoids redistribute themselves during the course of an interaction. This is explored further in Chapter 4.

Pseudointerference

Mutual interference causes a density-dependent reduction in the per capita searching efficiency as parasitoid density increases, as already seen in Fig. 2.8. Interestingly, the same kinds of relationship can arise without any behavioural

(a)

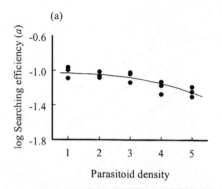

(b)

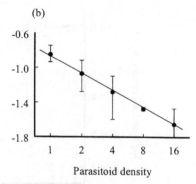

Fig. 2.11 Interference relationships for *Venturia canescens* parasitising larvae of the flour moth, *Ephestia cautella*. (a) Results from a laboratory experiment in which 564 hosts were distributed in a thin layer of wheat toppings on the floor of a cage of 50×50 cm. (b) As (a) but the host larvae are now confined to 16 small containers on the floor of the cage. (After Hassell 1978.)

interference on the part of the searching parasitoids, simply because *all* processes causing a density-dependent reduction in parasitoid searching efficiency (measured from eqn (2.5)) cause an apparent interference relationship. The negative binomial model (2.7) provides a good example of this: for small values of k the aggregated distribution of parasitism amongst hosts causes the average searching efficiency across the whole parasitoid population to decline with increases in parasitoid density (Fig. 2.12). This happens because an increasing proportion of encounters are wasted on hosts that have already been parasitised compared to the outcome with random search (Taylor 1993b). The stabilising effects of small values of k can thus equally well be represented by the value of m^* as described in the previous section.

Such apparent interference arising from aggregated, rather than random, distributions of parasitism has been dubbed 'pseudointerference' by Free *et al.* (1977) (see also Stinner and Lucas 1976; Beddington *et al.* 1978; Münster-Swendson and Nachman 1978). It arises whenever the risk of parasitism varies between host individuals, whatever the cause, rather than from parasitoids directly interfering with each other. Indeed, it is only when displayed in this one particular way (Fig. 2.12) that the connection with interference is made. Pseudointerference is therefore central to understanding the importance of non-random parasitism, which is discussed in Chapter 4, but as a term it is rather misleading.

Indirect mutual interference

Another category of interference has been introduced by Visser and Driessen (1991, 1999). They define *indirect mutual interference* as a 'behaviourally caused decrease in searching efficiency [that] can only be detected at the population

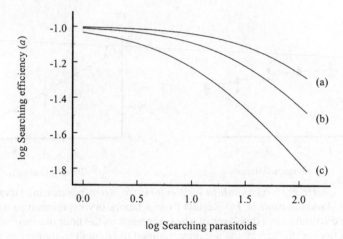

Fig. 2.12 Apparent interference relationships ('pseudointerference') arising solely from the aggregated distribution of parasitoid attacks on hosts generated from the negative binomial model (2.7). In each case $a = 1$; and in (a) $k = 5$, (b) $k = 2$ and (c) $k = 0.6$.

level'. This idea stemmed from a laboratory experiment by Visser *et al.* (1990) in which parasitoids (*Leptopilina heterotoma*) searched for hosts (*Drosophila* spp.) confined to a number of discrete patches. Although no mutual interference was seen at the scale of a single patch, Visser and Driessen (1991) claimed that interference was still occurring at the population level when searching efficiency was averaged across all the patches, *without* pseudointerference being in any way involved. An application of this in the field has been given by Cronin and Strong (1993).

Weisser *et al.* (1997), however, disagree that this category of interference represents something fundamentally different. They argue that the results of Visser and Driessen's model can easily be understood in terms of mutual and pseudointerference, and that proliferating the terms is therefore unnecessary. Indeed, both categories would have been unambiguously seen had Visser and Driessen changed parasitoid density in their model (Fig. 3 in Weisser *et al.* 1997). Driessen and Visser (1997) counter this by adopting a behaviourally based definition of pseudointerference in terms of an unequal time spent on patches by parasitoids, instead of the more usual concept of pseudointerference depending on variation in the risk of parasitism. This leads them to conclude, for example, that the negative binomial model (2.7) of May, which is rendered stable *solely* by such heterogeneity of risk (see Chapter 4), combines not only pseudointerference but also their category of indirect mutual interference (see also Lynch 1998). The problems seem to stem from a behavioural definition for pseudointerference that cuts across Free *et al.*'s (1977) original definition of a purely population dynamical phenomenon expressing the stabilising effects of aggregated distributions of parasitism amongst the host population as a whole. A clear discussion of this debate has recently been given by Lynch (1998), who defines Visser and Driessen's indirect mutual interference very succinctly in terms of changes in the degree of parasitoid aggregation in a patchy habitat as parasitoid density changes. Important though this phenomenon may be (see Chapter 4), it still remains the case that such spatial effects can be captured within the original definition of pseudointerference.

Setting these disagreements aside, it is clear that disentangling the different components that contribute to these patterns of density-dependent decline in searching efficiency with parasitoid density will often be difficult, and can only be done with detailed data on parasitoid behaviour and on the distribution of parasitism amongst the host individuals.

2.4.3 Ratio-dependence

The obvious step of combining functional responses with interference by varying both host and parasitoid densities has a long-standing tradition (e.g. Watt 1959; Hassell and Rogers 1972; Hassell and May 1973; Free *et al.* 1977; Sutherland 1983; Ruxton *et al.* 1992; Moody and Ruxton 1996), leading to models with considerably increased flexibility. For example, combining the linear interference in eqn (2.9) with a Holling type II functional response for parasitoids (see row K in Table 2.1) produces models of intermediate stability. Thus the stable region in Fig. 2.9 progressively shrinks as the proportion of time spent handling is increased (Hassell and May 1973).

A less mechanistic, but simpler, way of combining the effects of parasitoid and host (or predator and prey) densities is in terms of their *ratios*. In the context of the host–parasitoid framework (2.1), the issue is whether the survival function $f(\cdot)$ is better given by $f = (N_t, P_t)$ or by $f = (N_t/P_t)$. Since the original proposition by Getz (1984), Ginzburg (1986) and Arditi and Ginzburg (1989), views have become quite polarised on the value of ratio-dependent responses in host–parasitoid and predator–prey models. The virtues of ratio-dependence have been widely extolled (e.g. Getz 1984; Arditi and Akçakaya 1990; Berryman 1992; Ginzburg and Akçakaya 1992; Slobodkin 1992; Akçakaya *et al.* 1995); and roundly criticised (Abrams 1994; Gleeson 1994).

The arguments range broadly. Here we focus on only one category of empirical evidence and argue, for parasitoids at least, in favour of the more mechanistic approach of explicitly combining host and parasitoid densities as free variables within the function $f(\cdot)$. Let us commence with a straightforward combination of the mutual interference in eqn (2.9) with Holling's disc equation (Row D, Table 2.1) to make a functional response to host density that is also dependent on the numbers of searching parasitoids, as illustrated by Row K in Table 2.1 (Rogers 1972; Hassell and May 1973; Free *et al.* 1977; Arditi and Akçakaya 1990). Interestingly, Arditi and Akçakaya (1990) point out that if $m = 1$ in this expression the functional response becomes dependent only on the ratio N_t/P_t rather than on the absolute values of each. In other words, parasitoid interference compensates perfectly for any variation in parasitoid density and the response therefore becomes purely ratio-dependent. In support of the generality of this, they argue that conventional methods of estimating the interference constant m from laboratory experiments of the kind shown in Fig. 2.8 have tended to underestimate the true value of m, particularly in those cases where parasitoids can at least partially discriminate between healthy and already-parasitised hosts, and also in the case of predators (where the discrimination between healthy and eaten prey is perfect!). The problem lies in the use of eqn (2.5), which assumes zero handling time, to calculate the values of searching efficiency at the different parasitoid or predator densities. When allowing for this, Arditi and Akçakaya found, in 15 sets of laboratory data they examined, that the revised estimate of m was in most cases no longer significantly different from unity. They thus claim empirical support for the use of ratio-dependent functional responses in general.

There are difficulties, however, with this interpretation. First, even if $m \approx 1$ was typical from laboratory experiments, these levels are likely to be greatly exaggerated by the small and confined experimental conditions. Second, the argument that m-values in the field are typically in the region of one is very difficult to sustain; for example, values of k in the negative binomial model above have to reach the limit of zero for m (now representing pseudointerference) to become one (Fig. 2.12). Finally, the condition $m = 1$ has unfortunate dynamical implications within model (2.1). When $T_h = 0$, it represents the boundary between parasitoid density dependence that is undercompensating (and hence stabilising) and parasitoid density that is overcompensating, enabling the host population to 'escape' from parasitism and increase exponentially with the parasitoid population never able to 'catch-up'. With finite handling times, this boundary is lowered so exacerbating the effect (Hassell and May 1973).

In short, these ratio-dependent, host–parasitoid models have rather pathological properties. A more mechanistic approach of including independently-varying host and parasitoid densities produces more flexible models in which the separate parameters determining the functional and interference responses can be identified, parameterised and their effect on dynamics quantified.

2.5 Summary

This chapter has reviewed three basic models for host–parasitoid interactions, all of which assume discrete generations and synchronised host and parasitoid life cycles. In the model of Thompson (1924), each parasitoid lays a constant number of eggs scattered randomly amongst the available hosts. The model is unstable; either both populations increase unchecked or the parasitoids drive the hosts to extinction. Nicholson (1933) and Nicholson and Bailey (1935) also assume random host encounters by each parasitoid in their model, but now encounters are in proportion to host abundance. Parasitoids are therefore search-limited rather than egg-limited, and thus have a linear functional response (whose slope is the searching efficiency). The model has an unstable equilibrium around which the populations show rapidly expanding oscillations. Finally, in May's (1978) model, there is the same linear functional response, but now the distribution of encounters with hosts is aggregated following a negative binomial distribution. In contrast to the other two, this model is stable if there is sufficient aggregation in this distribution of attacks (i.e. k of the negative binomial must be less than one).

There have been many elaborations on these basic assumptions. For example, functional responses have been made more realistic by limiting the parasitoid's maximum attack rate at high host densities. Type I functional responses rise linearly with increasing host density to an abrupt plateau, type II responses include a handling time and therefore rise asymptotically to a maximum level, and type III responses initially show an accelerating attack rate as host density increases before levelling off due to the influence of handling time (or egg-limitation). Further refinements include those in which there are different handling times depending on whether the host has already been parasitised or not, and in which the parasitoids can either always be egg-limited (Thompson) or search-limited (Nicholson), or any shade in between.

Other elaborations have involved interference between searching adult parasitoids. This may result from direct behavioural interactions between searching females (mutual interference), or from non-random distributions of attacks on hosts (pseudointerference). Because interference involves a density-dependent reduction in parasitoid efficiency as parasitoid density rises, it can strongly promote population stability. But because the frequency of encounters between parasitoids under natural conditions will usually be rather low, mutual interference is unlikely to be generally important to host population dynamics. Pseudointerference, on the other hand, arises from any process that generates non-random parasitism and should therefore have a greater impact on the populations; this is fully discussed in Chapter 4. Finally, there have been attempts to combine the effects of both host and parasitoid densities within a single framework, in which it is the ratio of the two populations that is all important. Although having the virtue of simplicity, the more mechanistic approach of separately, and explicitly, incorporating the effects of host and parasitoid densities gives a better understanding of the underlying dynamics.

Notes

1. By ovipositing beside or under the scale where the larvae then feed, the female beetles are acting much as parasitoids.

2. Other systems with continuous birth–death processes may also be approximated by discrete models, depending on the internal dynamics forcing the system to show cycles of approximately one generation period (see Chapter 5).

3. This second term corresponds to the requirement that, at equilibrium, parasitism must 'remove' any host fecundity over and above that needed for replacement.

4. The laying of one or more eggs by a parasitoid on a host that has already been parasitised by the same species.

5. For one particular set of conditions ($w = \lambda \ln(\lambda)/(\lambda - 1)$ and $N_1/P_1 = \lambda/(\lambda - 1)$) an infinite number of host–parasitoid equilibria exist.

6. Pro-ovigenic parasitoids emerge with their full complement of eggs, while synovigenic parasitoids continually mature eggs throughout their adult lives, and therefore usually host-feed to provide the necessary nutrition.

7. Handling time (see p. 21) is therefore implicitly zero.

8. Throughout the book the abbreviation 'ln' is used for $\log_e$ and 'log' is used for $\log_{10}$.

9. Notice that a has units of *area* $\times$ *time*$^{-1}$ if hosts and parasitoids are expressed as densities per unit area, and units of *time*$^{-1}$ if the populations are expressed as dimensionless numbers.

10. In a broadly parallel development, the negative binomial distribution has been used to describe the distribution of true parasite infections in hosts (e.g. Anderson 1978; Anderson and May 1978; Hudson *et al.* 1992) and forms the basis of several models of host–parasite and host–pathogen interactions (e.g. Crofton 1971; Anderson and May 1978; May and Anderson 1978; Dobson and Hudson 1992).

11. Note that the 'predator' response will also apply to parasitoids that unfailingly and instantly detect hosts that are already parasitised (and therefore spend no time handling them).

12. Interactions between the effects of a type II functional response and density-dependent resource limitation of the host are considered in Chapter 3.

13. Note that eqn (2.8) is dimensionally correct only when P represents the total population of parasitoids rather than a density; otherwise a better form is based on $a = Q(QP_t)^{-m}$ (Beddington *et al.* 1978).

3

Parasitism and host density dependence

3.1 Introduction

The host–parasitoid models described in Chapter 2 have four basic features: discrete and synchronised generations; a description of host survival from parasitism, $f(\cdot)$; a host rate of increase, λ; and a further parameter, c, converting hosts parasitised in generation t to searching parasitoid females in the following generation $t + 1$. Together these embrace a huge amount of host and parasitoid biology. Thus the function $f(\cdot)$ includes everything affecting the success of the searching parasitoids in finding and parasitising the N_t hosts. The host rate of increase, λ, is not just the average fecundity per adult host, but is the net rate of increase after discounting all the various mortality factors affecting the host population other than parasitism. For insects, a typical fecundity of the order of 100 or so eggs per female can thus easily translate to a value of λ of less than 10. Also, if any of these components are density-dependent then λ should be a function of host density. In a similar vein, the term c encompasses a large part of the parasitoid's life table and depends on everything affecting the survival and sex ratio of parasitoid progeny from the time of parasitism to the time of search by the next generation of female parasitoids.

In this chapter, rather than keep λ and c constant as was the case in Chapter 2, we first examine the effects of making c density-dependent, either via parasitoid survival or via shifts in sex ratios, and then consider the impact of direct density dependence acting on the hosts, either by making λ density-dependent or by introducing a density-dependent host mortality caused by generalist natural enemies.

3.2 Parasitoid survival and sex ratios

Most host–parasitoid models assume that the number of searching-female parasitoids stems directly from the number of hosts parasitised in the previous generation (i.e. $c = 1$ in model (2.1)). This is very convenient, but strictly corresponds only to parthenogenetic parasitoid species in which a solitary larva always survives from each parasitised host. Real parasitoids may have shifting sex ratios and gregarious rather than solitary larvae. Parasitoid larvae may also suffer a range of mortalities; for example, they may be encapsulated by the host's haemocytes (Salt 1963, 1970; Nappi 1975), or they may perish if the parasitised hosts die prematurely from predation, disease or other causes. Any of these would mean that $c \neq 1$.

Parasitoid species also vary considerably in the average number of eggs laid in or on a host at the time of parasitism.[1] While the majority of parasitoids have solitary larvae (Godfray 1994), in which only one parasitoid progeny emerges from each host parasitised irrespective of how often that host has been encountered, there are many species with gregarious larvae in which the adult females lay clutches of eggs on or in each host attacked. The number of offspring emerging from a parasitised host is now more variable, and depends on the number and size of the clutches laid on the host and on the survival of the parasitoid larvae. In most gregarious species an important influence on this survival is competition between the larvae for the limited host resource. This affects larval survival to some extent, but in particular influences the body size of the subsequent adult females and hence their fecundity (Benson 1974; Waage and Lane 1984; Waage and Ng 1984; Taylor 1988a; Hardy *et al.* 1992; Godfray 1994; Visser 1994; Kazmer and Luck 1995; West *et al.* 1996). The dynamical consequences of this are straightforward, as long as each parasitised host tends to yield a constant number of similar-sized female parasitoids for the next generation. The parameter, c, will then be a constant greater than one, which will only have a scaling effect on the host equilibrium levels. But more usually in such cases, the number and quality of adult parasitoids emerging from a host is likely to depend on both the number of parasitoid larvae initially present and on the extent of within-host competition between the feeding larvae, thereby introducing density dependence.

A general framework to show the dynamical importance of within-host parasitoid competition has been developed by Taylor (1988b). He considers the following possibilities, assuming throughout that a constant number of parasitoid eggs are laid at each host encounter (Fig. 3.1). If there is no within-host competition, each successive encounter with a host will result in further surviving parasitoids, as shown in line A. Lines B–D show the effects of different degrees of competition between the parasitoid larvae. This may be *under-compensating* (line B) producing a monotonically increasing yield of surviving parasitoid progeny as the number of competing parasitoid larvae per host increases; *perfectly compensating* (line C) corresponding to a pure 'contest' competition in which a constant number of parasitoid progeny survive irrespective of the number of times a host is encountered; or *overcompensating* (line D) corresponding to 'scramble' competition in which the survival of parasitoid larvae falls at high densities (see p. 40 for a discussion of contest and scramble). The sequence A–D therefore represents a series of increasing strengths of density dependence acting on the parasitoid progeny.

One of the cases in (Fig. 3.1) is already familiar: all the models considered in Chapter 2 have assumed line C where the parasitoid survival term c is a con-

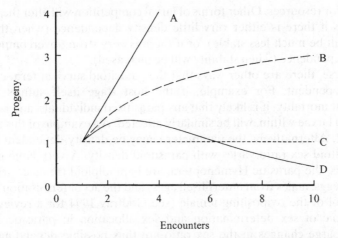

Fig. 3.1 Some effects of within-host competition on the survival of parasitoid progeny in relation to the number of times a host is encountered by a searching parasitoid (assuming a constant number of parasitoid eggs are laid at each encounter). The four lines A to D represent a series of increasing strengths of density dependence acting on the parasitoid larvae within a host. (A) No within-host competition at all—each successive encounter with a host results in a further surviving parasitoid; (B) *under-compensating* density-dependent competition; (C) *perfectly compensating* density dependence corresponding to a pure contest competition; (D) *overcompensating*, scramble competition. See text for a discussion of the stability implications of these different degrees of density dependence. (After Taylor 1988*b*.)

stant (equal to one). These models therefore implicitly contain strong density-dependent contest competition acting on the parasitoid population. To illustrate the effects of altering this assumption, Taylor (1988*b*) takes model (2.7) with a negative binomial distribution of attacks and a constant host rate of increase, which for constant c is stable for all $k < 1$ (May 1978). This therefore provides a convenient 'baseline' from which to compare the effects of different degrees of density-dependent parasitoid competition. Starting with line A, in which there is no larval competition and therefore no density dependence acting on the parasitoid population at all, we note that stability is impossible in this case. If we now introduce some weak (i.e. undercompensating) density dependence (line B) the interaction can be stable as long as k is small enough, but it is not as stable as the baseline model (May 1978) with constant c. Finally, when the density dependence at equilibrium is much stronger and overcompensating (line D), stability is increased and can occur for much less-aggregated distributions of attack (i.e. for values of k considerably above one). In short, implicit density dependence acting on the parasitoid larvae is crucial to the dynamics of host–parasitoid interactions. The conventional host–parasitoid models described in Chapter 2 apply either to parasitoids with solitary larvae or to gregarious species in which the larvae 'contest' for resources. Other forms of larval competition will alter the stability properties if there is either very little density dependence (when the inter-actions will be much less stable), or if there is very strong overcompensating density dependence (when stability will be increased).

Of course, there are other ways that the parasitoid survival term c may be density-dependent. For example, if the host stage itself suffers density-dependent mortality, it is likely that any parasitised individuals, and hence the parasitoid larvae within, will be similarly affected. An example of this is shown in Fig. 3.2. Alternatively, the parameter c may be density-dependent because the parasitoid sex ratio varies with parasitoid density. A very large group of parasitoids, the parasitic Hymenoptera, are haplodiploid (females arise from fertilised eggs, males from unfertilised eggs) and the act of fertilisation is under the control of the ovipositing female (see Godfray 1994 for a review of the whole topic of sex determination and sex allocation in parasitic Hymen-optera). Large changes in the sex ratio are thus possible depending on the ways that females respond to surrounding conditions. Pronounced density-dependent shifts in sex ratios, leading to increasing proportions of male progeny as the density of searching parasitoids rises, have been frequently observed in the laboratory (Wylie 1966; Walker 1967; Werren 1983; Waage and Lane 1984; Mohamed and Coppel 1986; Strand 1988; Godfray 1994; see

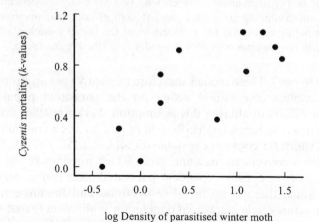

log Density of parasitised winter moth

Fig. 3.2 Density-dependent larval and pupal mortality of *Cyzenis albicans*, a tachinid parasitoid of the winter moth, *Operophtera brumata*, in relation to the density (m^{-2} canopy area) of parasitised winter moths. The mortality is largely ascribed to invertebrate (carabid and staphylinid beetles) and vertebrate (shrews) predators eating parasitised winter moth pupae in the soil. Mortality is expressed as k-values (Varley and Gradwell 1960), calculated from the difference between the log number of winter moth larvae and prepupae containing *Cyzenis* that fall from 1 m^2 of canopy area and the subsequent log number of adult *Cyzenis* per m^2 emerging the following year (Hassell 1969b). The mortality closely mirrors the density-dependent mortality of winter moth pupae in the soil shown in Fig. 3.7(c).

also Ueno (1999) for host size-dependent shifts in parasitoid sex ratios), and occasionally in the field for parasitoids of gall wasps (Hails 1989; Hails and Crawley 1992) and in pollinating fig wasps (Frank 1985; Herre 1985, 1987).[2]

The dynamical implications of density-dependent sex ratios have been explored by Hassell *et al.* (1983). Briefly, density-dependent sex ratios contribute to population stability; they can even stabilise the Nicholson–Bailey model if λ is small enough, and the sex ratio shifts large enough. However, just as with the laboratory experiments on mutual interference (see p. 28), the high levels of crowding within laboratory cages are likely to exaggerate the effect compared to natural conditions, where any density-dependent shifts in sex ratio are more likely to arise from local mate competition (Hamilton 1967) rather than from direct crowding. This would tend to restrict the effect to a rather narrow range of parasitoid densities, and make it less important dynamically.

3.3 Parasitism and host density dependence

Simple host–parasitoid models often assume a constant host rate of increase. In this way the natural enemy's impact on dynamics can be more readily discerned, without the complication of additional factors affecting the host population. The assumption is more likely to be valid if the parasitoid population characteristically depresses host abundance well below its parasitoid-free carrying capacity (Beddington *et al.* 1978), in which case any density-dependent competition for resources by the host will be negligible. Opinions differ amongst ecologists on the relative importance of such 'top-down' regulation (by natural enemies) compared to 'bottom-up' regulation (by resource limitation) (see Chapter 2.1), but few would disagree that examples of both, as well as the full range of intermediates, are bound to occur. The remainder of this chapter is concerned with the interplay of these processes. We first examine some simple models of direct density dependence acting on a host population in the absence, and then in the presence, of parasitism. Host–parasitoid interactions in which the host density dependence is caused by generalist natural enemies are then described. Such host–specialist/generalist interactions pave the way for the multispecies host–parasitoid systems discussed in Chapter 6.

3.3.1 Host density dependence on its own

In contrast to the difficulties of detecting density dependence in the field, it has been easy to demonstrate and quantify density dependence from laboratory experiments with insects. A range of densities are typically kept with a fixed amount of resource for a set period, and the numbers surviving (before or after reproduction) scored. Any density dependence is then easily revealed (e.g. Nicholson 1954; Bakker 1961; Bellows 1981, 1982a,b). The examples in Fig. 3.3

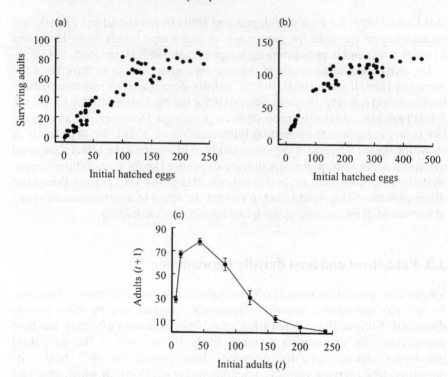

Fig. 3.3 Examples of density-dependent competition for resources from insect laboratory experiments. (a) Density-dependent relationship between the number of hatched eggs and the number of surviving adults of the bruchid beetle, *Callosobruchus maculatus*. (b) As (a), but now involving *Callosobruchus chinensis*. (c) As (a), but showing mean numbers of adult beetles in successive generations, t and $t + 1$ ($\pm$ SE). (After Bellows 1982a.)

illustrate three fundamental categories into which such density dependence can fall. The number of survivors may tend to be undercompensating (Fig. 3.3(a)), may rise monotonically to some more-or-less constant value as population density increases (perfectly compensating) (Fig. 3.3(b)), or the relationship may be dome-shaped (Fig. 3.3(c)) in which case survivors rise to a peak at intermediate population densities, above which overcompensation occurs. These contrasting patterns can be generated by two, fundamentally different, kinds of intraspecific competition: contest competition and scramble competition[3] (Nicholson 1954). Contest involves individuals striving to secure their own share of the resource at the expense of others. Territorial species are obvious examples, as are solitary insect parasitoids where, irrespective of how many eggs are laid in a host, only a single adult parasitoid subsequently emerges. In a contest, there tend to be clear winners and clear losers, and the upper asymptote in Fig. 3.3(b) is set by the maximum number of such winners that the resource can support. In scramble competition, the resources are shared much more equally between the competing individuals, and hence the ill effects is also more equally spread amongst the competitors. Between these two cases are a continuum of intermediates; for instance, there may be a contest process for a fixed number of territories, but these territories may tend to shrink in size as the density of competitors increases (e.g. Watson and Miller 1970).

The effects of the density-dependent patterns in Fig. 3.3 on the dynamics of simple models of population growth are easily portrayed. Consider an insect population with discrete generations and no natural enemies:

$$N_{t+1} = \lambda N_t g(N_t) \tag{3.1}$$

where λ is the finite rate of increase per adult and $g(N_t)$ either describes the density-dependent survival of the N_t insects or, as $\lambda g(N_t)$, the density-dependent natality. Many alternative expressions for g have been proposed (Table 3.1). All can be pigeonholed as being under-, over- or exactly compensating, or having the flexibility to be any of these. Their dynamical effects within eqn (3.1) are well known: depending on the parameters of population growth and density dependence, the population moves either to a stable point, shows limit cycles or is chaotic (May 1974*a*; May and Oster 1976).

In this section, we examine the effects of such simple density dependence on a single species system, but now set in an explicitly patchy environment. Let us consider a habitat containing discrete patches of foodplants for the larvae of an herbivorous insect population. The insects have discrete generations, and in each generation the adult females disperse amongst patches laying their eggs on the plants. The emerging larvae, and subsequently the pupae, complete their development within their natal patch prior to the next generation of adults emerging. If all patches were identical in the number of insects they contain, and all the insects had the same probability of survival, then the

Table 3.1 Some simple functions for density dependence of the form of eqn (3.1). Those containing the parameter b may be undercompensating ($b < 1$), perfectly compensating ($b = 1$) or overcompensating ($b > 1$). (After May and Oster (1976) and Bellows (1981.)

Entry	$g(N)$	Author(s)
A	aN^{-b}	Varley and Gradwell (1963*b*); Varley *et al.* (1973)
B	$\exp(-aN)$	Moran (1950); Ricker (1954); Cook (1965); May (1974*b*); and others
C	$(1 + aN)^{-1}$	Skellam (1951); Pielou (1969)
D	$\exp(-aN^b)$	Bellows (1981)
E	$[1 + (aN)^b]^{-1}$	Maynard Smith and Slatkin (1973)
F	$(1 + aN)^{-b}$	Bleasdale and Nelder (1960); Hassell (1975); Hassell *et al.* (1976*b*)

spatial structure of the environment would be largely irrelevant. Uneven distributions of the insects, however, whether due to variations in patch 'quality' or other processes, introduce important differences between the comparable homogeneous and heterogeneous systems. Birch (1971) describes in detail such a system involving the cactus-feeding moth, *Cactoblastis cactorum*, feeding on prickly pears (*Opuntia* spp.). The distribution of *Cactoblastis* egg sticks per plant is highly aggregated, which increases the chance of both cactus and *Cactoblastis* populations surviving. Were the eggs to be distributed evenly the plants would be more likely to be uniformly destroyed once the population density rose high enough.

Patchy, single-species systems of this kind have been explicitly modelled by DeJong (1979, 1981). She assumed a habitat containing n patches, adults that lay a fixed complement of F eggs, and a within-patch, density-dependent survival of the immature stages depending on the population density per patch. Two kinds of dispersal between plants were considered:

1. In her 'adult dispersal', the mated adult females that emerge from the different patches enter a dispersal phase during which they mixed thoroughly, before going on to search for plants. The resulting distribution of adults from plant to plant was assumed, in turn, to be either uniform, binomial, Poisson or negative binomial. Once on a plant, each female then lays her full complement of eggs; thus, only groups of F eggs can occur on a plant, and an egg never finds itself in a density lower than this clutch size.
2. DeJong's so-called 'juvenile dispersal' could occur in two ways: either by the population of immature stages actively dispersing between plants or, as is more likely, by the adult females flitting from plant to plant and laying a single egg at each encounter with a plant. In this second case, it is the egg distribution, rather than the adult females, that is assumed to be either uniform, binomial, Poisson or negative binomial, and now the possible egg densities per plant are 0, 1, 2, 3, ..., instead of the 0, F, $2F$, $3F$, ... from the 'adult dispersal' case.

These two scenarios lead to much the same model, but there are differences in the number of patches occupied by the insects (greater for juvenile dispersal) and in the variance in egg densities per patch (greater for adult dispersal, particularly at low total population densities).

A model of these spatially distributed populations can be portrayed as follows, taking the juvenile dispersal case as an example (DeJong 1979):

$$J_{t+1} = nF\sum_{i=0}^{\infty}[ip(i)g(i)] \tag{3.2}$$

where J is the total population size of juveniles (e.g. eggs) over all patches. The summation represents the average number of surviving offspring emerging from any one of the n patches, $p(i)$ is the probability of having i eggs in a patch (described by one of the four probability density functions described above), and $g(i)$ is a function giving the density-dependent probability of an egg surviving to an adult on a plant where i eggs have been laid. DeJong chose the density-dependent function, $\exp(-di)$ for this survival (entry B in Table 3.1), although alternative forms of density dependence could equally well have been used (Hassell and May 1985).

The key components of this model that affect dynamics are:

(1) the spatial distribution of the individuals;
(2) the severity of the within-patch density dependence, d; and
(3) the adult fecundity, F.

Of the four different spatial distributions considered by DeJong, the uniform distribution is the most unlikely and the easiest dealt with: dividing the population equally amongst n patches has no effect on the dynamics compared with the corresponding homogeneous case. The remaining three distributions, the binomial, Poisson and negative binomial, all involve some variation in the number of individuals from patch to patch, and this variation, if strong enough, can have a profound influence on the equilibrium and stability properties of the population. Let us consider in particular the case of the negative binomial distribution. As described in Chapter 2, it is defined by two parameters: the mean of the distribution and a parameter, k, that inversely defines the extent of clumping (most aggregated as $k \to 0$, becoming random or Poisson when $k \to \infty$). The stability properties are shown in Fig. 3.4. Stability is enhanced in

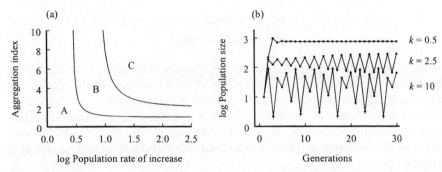

Fig. 3.4 Stability properties of the DeJong model (3.2). (a) Boundaries separating regions of different dynamical behaviour in terms of the degree of aggregation (expressed as $k + 1$) and the log rate of increase of the population as a whole, $\lambda = F\exp(-dF)$. The model is locally stable within region A (where there is monotonic damping) and B (where there are damped oscillations). In region C there are limit cycles and chaotic behaviour. (b) Three numerical examples illustrating the three regions in (a) with k as shown and $\lambda = 25$. Note the increasing stability as the degree of aggregation is increased. Interestingly, and fortuitously, the model collapses exactly to a much simpler model for density-dependent population growth (Entry F in Table 3.1) if the 'microscopic' parameters in eqn (3.2) are redefined in terms of three 'macroscopic' parameters as follows: $\lambda = F\exp(-dF)$, $a = [1 -\exp(-df)]/nk$ and $b = k + 1$ (Hassell 1980*b*; Hassell and May 1985).

two ways: (1) by increasing aggregation (small values of k), because this accentuates the degree of within-patch density dependence acting on the immature stages; and (2) as in most population models, by low overall rates of increase of the population—which depends on both the fecundity F and the extent of the within-patch density-dependent survival of the immature stages, d.

DeJong's model has been outlined in some detail as an example of how explicit patchiness may be introduced into simple population models, an approach that we return to in the next chapter. Although primarily a heuristic tool for exploring how spatial aggregation can interact with other demographic parameters affecting population dynamics, the model can also be extended to represent the dynamics of some real laboratory and field populations (1987; Hassell *et al.* 1989; Southwood *et al.* 1989)

3.3.2 Parasitism and host density dependence combined

Nicholson (1933) and Nicholson and Bailey (1935) suggested that one way in which the instability of their basic host–parasitoid model could be overcome was by the populations existing as metapopulations, in which local instability is countered by asynchrony in the extinctions and colonisations (Chapter 7). Another, more obvious, way of stabilising their models is to introduce further density dependence acting on the host population via mortality or natality. A simple representation of this is given by:

$$N_{t+1} = \lambda N_t g(N_t) f(P_t)$$
$$P_{t+1} = N_t [1 - f(P_t)] \tag{3.3}$$

where the function $g(N_t)$ represents the density dependence acting on the hosts and $f(P_t)$ is the familiar function for survival from parasitism. Varley and Gradwell (1963*a*) were the first to explore a version of (3.3) in which the density dependence, in the form of entry A in Table 3.1, stemmed from their data on pupal predation of the winter moth (see p. 55). They illustrated how the unstable Nicholson–Bailey model could be stabilised by sufficiently high values of b.

The inclusion of density dependence acting on the host population in host–parasitoid models introduces a number of possibilities. The parasitoids may be unable to maintain themselves and hence go extinct. Equivalently, if the parasitoids are absent they may be unable to invade the system in the first place. The conditions for such invasion from eqns (3.3) are straightforward: if K is the host equilibrium in the absence of the parasitoids (given by $\lambda g(K) = 1$) and parasitism is a function only of searching parasitoid density (i.e. there is a linear functional response), then the parasitoids will be able to invade the host population as long as the searching efficiency, a, is large enough in relation to K; in particular, if $a > 1/K$ (May and Hassell 1988; Kaitala *et al.* 1999).[4] If the parasitoids *can* persist in the system, the dynamics are then some amalgam of the effects of the host density dependence and the parasitoids, as discussed below.

The first thorough treatment of the way that parasitism and host density dependence interact was by Beddington *et al.* (1975). They assumed Nicholson–Bailey parasitoids and a host rate of increase in the form of a discrete version of the logistic equation, $g(N_t) = \exp(-aN_t)$, where $a = \ln\lambda/K$ and K is the carrying capacity in the absence of parasitism (May 1974*b*). They also introduced a very useful way of displaying the stability of the interactions by defining a parameter, q, as the proportional extent to which the parasitoids depress the host equilibrium, N^*, below K (i.e. $q = N^*/K$), and plotting this against the host intrinsic rate of increase, $r = \ln(\lambda)$ (Fig. 3.5(a)). Thus when q is close to one, the parasitoids are hardly depressing the host equilibrium at all; but as q decreases, the parasitoids have an increasing effect in reducing the host equilibrium further and further from the parasitoid-free carrying capacity, K. On these axes, invasion is possible for all $q < 1$, but the interaction is only stable within the area B in Fig. 3.5(a). Since the Nicholson–Bailey model is always unstable, this stable region arises solely from the direct density dependence acting on the host. Outside this, in the unstable region A, the host density dependence still dominates, but because it is overcompensating and the host's rate of increase is sufficiently high, limit cycles and ultimately chaos occur. The instability in the lower region C is completely different. The parasitoids are now driving the dynamics by having a high searching efficiency relative to K; hence, the host equilibrium is too far below K for the density dependence to have much effect and expanding host–parasitoid oscillations occur. In short, a Nicholson–Bailey model can be stabilised by the addition of host density dependence, but only if the parasitoids are relatively inefficient and the host equilibrium is consequently relatively close to K (Beddington *et al.* 1978).[5]

This picture changes somewhat when parasitism becomes a stabilising process; for instance, if described by the negative binomial model (2.7) (i.e. $f(P_t) = (1 + aP_t/k)^{-k}$). The invasion criterion above is unaltered but now, as long as $k < 1$, the parasitoids can stabilise the interaction without the help of any host density dependence (May *et al.* 1981). Hence, as the parasitoids progressively depress the host's equilibrium, the interaction is first primarily affected by the host density dependence and then increasingly stabilised by the parasitoids (Fig. 3.5(b)). What was previously an unstable region C is now a stable region with a largely parasitoid-maintained host equilibrium. In region A the parasitoids have little effect and the instability is driven, as before, by the host's strongly overcompensating density dependence. Therefore, once again the relative contribution of these two components—parasitism and the host density dependence—depends on the degree to which the parasitoid-maintained host equilibrium lies below the level of the host's carrying capacity. The precise stable areas in Fig. 3.5 depend on a linear functional response. They progressively decrease as the functional response becomes

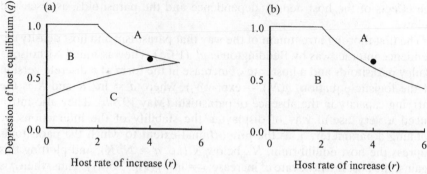

Fig. 3.5 Local stability boundaries for Model (3.3) with $g(N_t) = \exp(-aN_t)$ as described in the text. The boundaries are expressed in terms of the depression of the host equilibrium ($q = N^*/K$, where N^* and K are the host equilibria with and without parasitoids, respectively) and the host rate of increase ($r = \ln\lambda$). The interactions are locally stable in region B, unstable due to overcompensating host density dependence in region A and unstable with expanding host–parasitoid cycles in region C. (a) Random parasitism, $k \to \infty$, and (b) aggregated parasitism with $k = 0.5$. The point shown on each figure is discussed in the text.

more strongly type II in shape and therefore increasingly influenced by a maximum attack rate; this has been thoroughly explored by Lane *et al.* (1999).

Apparently anomalous results can be resolved with this analysis in mind. For example, Hochberg and Lawton (1990) comment on the counterintuitive way that increasing the aggregation of parasitoid attacks (specifically, reducing k in the negative binomial model), which one would normally expect to increase stability, can sometimes actually *decrease* stability in models with host density dependence. This arises straightforwardly from the way that small values of k also increase the host equilibrium level (see p. 19). Small values of k therefore move the host population towards the region where the influence of the host density dependence predominates, and which may drive higher order behaviour. For example, the point marked in Fig. 3.5 falls within the stable region in Fig. 3.5(a) where the parasitoids are attacking randomly ($k \to \infty$), but outside it in Fig. 3.5(b) when $k = 0.5$. In other words, aggregation of attacks has moved the host equilibrium out of the stable region and into the unstable region where the cycles are driven by the host's density dependence.

There are other ways that parasitism and host density dependence can interplay to produce interesting results. For instance, within any discrete-generation interaction there can be quite different dynamics depending on the ordering of the various events in the host's life cycle (Crawley 1975; Wang and Gutierrez 1980; May and Hassell 1981; May *et al.* 1981). Model (3.3) makes some quite specific assumptions about this: parasitism is assumed to act first in the host's life cycle, followed by the density dependence acting on the survivors from parasitism, but at a level set by the *total* hosts (N_t), parasitised or not. Such a

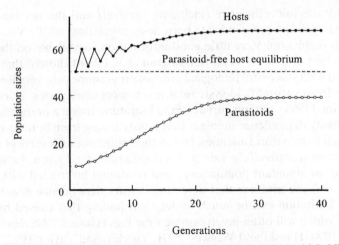

Fig. 3.6 Numerical simulations from a host–parasitoid model in which Nicholson–Bailey parasitism acts first in the host life cycle followed by overcompensating density dependence acting at a later stage (see text), where the host rate of increase, $r = 2$, its carrying capacity, $K = 50$, and the parasitoid searching efficiency, $a = 0.02$. In the absence of parasitism the host population is regulated at $N^* = 50$ (broken line). Following introduction of parasitoids (hollow circles), the host population (solid circles) rises due to overcompensation by the host density dependence. (After May *et al.* 1981).

situation could arise for example where the density dependence was caused by predation rather than resource limitation (see below), and the predators responded to the total density of their insect prey irrespective of whether the prey were parasitised or not.[6] Other, more obvious, scenarios include: (1) the density dependence may act first in the host's life cycle followed by parasitism at a later stage, in which case the parasitoid equation in model (3.3) becomes: $P_{t+1} = N_t g(N_t)[1 - f(P_t)]$; or (2) the density dependence may follow parasitism but now as a function only of the healthy hosts present at the time, in which case the host equation in model (3.3) becomes: $N_{t+1} = \lambda N_t g(N_t f(P_t)) f(P_t)$. The dynamics with (1) are little different from those shown in Fig. 3.5, but the dynamics of model (2) can be very different. In particular, since the density dependence follows the parasitism in the host life cycle, the addition of parasitism can actually raise the host equilibrium *above* the parasitoid-free level (i.e. $q > 1$) as long as the density dependence is overcompensating and the parasitoid efficiency rather low. An example is illustrated in Fig. 3.6.

Understanding the relative contributions of host density dependence and parasitism is clearly central to the debate (introduced in Chapter 2) on the frequency of 'top-down' or 'bottom-up' regulation of natural communities of insect herbivores. In principle, of course, either can apply depending on the system, or both may operate within the same system but act on different developmental stages (Kato 1994). The host equilibrium depends on the balance between parasitoid efficiency (including survival) and the net host rate of increase; large values of q indicate bottom-up regulation, small values indicate top-down regulation. Very little good information is available on the characteristic values of q in the field. Beddington *et al.* (1978) showed that q-values associated with successful biological-control programmes are very low, but the evidence from natural field systems is much more ambiguous. Harrison and Cappuccino (1995), in their survey of the literature, found a preponderance of direct density dependence in insect herbivores arising from bottom-up effects rather than from natural enemies. But as they point out such surveys are likely to be unrepresentative (e.g. rare insect populations tend not to be studied as frequently as abundant populations) and so should be treated with caution. There is also the problem that detecting density dependence resulting from resource limitation can be much easier than finding that caused by natural enemies, which will often involve some time lags (Hassell 1985; Hanski 1990; Turchin 1990; Hanski and Woiwod 1991; Turchin and Taylor 1992; Holyoak 1993, 1994*a,b*; Turchin 1995). In short, it is hard to be prescriptive and we should expect the full range of q-values to occur.

We have so far assumed that the density dependence acting on the host population, g in model (3.3), arises from competition for resources. But, equally, it could arise from other factors and especially from generalist natural enemies.

3.3.3 Generalist parasitoids and predators

Insects are typically attacked by a range of generalist (= polyphagous) natural enemies: either parasitoids, or predators (both vertebrate and invertebrate). By virtue of their broad diet, the reproductive rate of these generalists tends to be at least partially uncoupled from the abundance of any one of their host or prey species (Crawley 1992). If we assume that the generalists have a constant population size from generation to generation, there are a number of contrasting patterns of mortality that we may expect. First, as originally assumed by Howard and Fiske (1911), each generalist may tend to attack a constant number of hosts or prey. This will cause an inverse density-dependent mortality from generation to generation. Alternatively, generalists, particularly more sessile ones, may tend to capture prey in simple proportion to the prey's abundance, giving a density-independent relationship. And third, they may act as density-dependent factors, as shown in Figs 3.7(a), (b), without the time lags that drive the characteristic oscillations of completely coupled host–parasitoid or predator–prey systems (see Figs 2.3(b), (c)). Such patterns could arise, for example, if the generalists tend to aggregate in local areas where a particular prey type is most abundant (Buckner 1964; Royama 1970; Hanski and Parviainen 1985; Gould *et al.* 1990), thereby creating a 'switching' effect (see Murdoch 1969) between elements of their diet depending on which are the most abundant in a generation. These examples depend on the generalist populations per generation being relatively constant in size. If, however, their

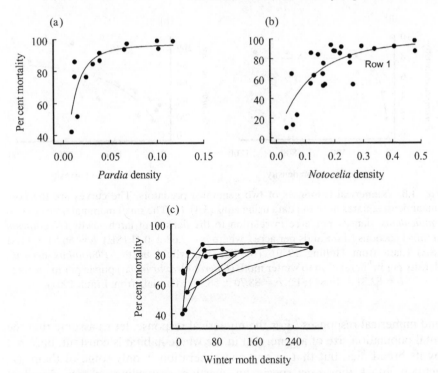

Fig. 3.7 Examples of density-dependent pupal mortality from generation to generation for three species of soil-pupating Lepidoptera, ascribed mainly to predation by carabid and staphylinid beetles. The curves in (a) and (b) are the non-linear, least squares fit of the data using eqns (3.4) and (3.5), with handling time (T_h) assumed to be zero. (a) Mortality of *Pardia tripunctatana* (per $0.18 \, m^2$) where $ah = 3.25 \pm 1.26$ (SE), $b = 0.03 \pm 0.02$ and $k \to \infty$ (data from Bauer 1985). (b) Mortality of *Notocelia roborana* (per $0.18 \, m^2$) where $ah = 9.37 \pm 7.99$ (SE), $b = 9.04 \pm 7.77$ and $k \to \infty$ (data from Bauer 1985). (c) Mortality of winter moth (*Operophtera brumata*) (per m^2) (data from Varley *et al.* 1973). The points are serially linked over the 18 generations forming an anticlockwise 'spiral' indicating a delayed density-dependent component to the mortality. In addition to predation by carabid and staphylinid beetles and small mammals, the mortality also includes parasitism by an ichneumonid parasitoid, *Cratichneumon culex*.

reproductive rate *is* influenced to some extent by the abundance of a particular host or prey type, we would expect there also to be a time-delayed component to the relationship, as shown in Fig. 3.7(c). This is intermediate between the direct density dependence shown in Figs 3.7(a), (b) and the delayed density dependence seen in Fig 2.3(c).

The density-dependent patterns shown in Figs 3.7(a), (b) can be described phenomenologically using the kinds of simple expressions in Table 3.1. But the impact of generalists on a particular host or prey population can also be represented more mechanistically by taking explicit account of their functional

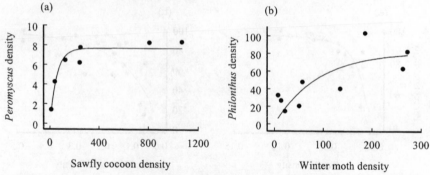

Fig. 3.8 Numerical responses of two generalist predators. The curves are the non-linear, least squares fit of the data using eqns (3.4). (a) The small mammal, *Peromyscus maniculatus*, density per acre in relation to the density of larch sawfly (*Neodiprion sertifer*) cocoons (thousands per acre), where $h = 7.68 \pm 0.41$ (SE), $b = 56.54 \pm 14.81$ (SE) (data from Holling 1959). (b) The staphylinid beetle, *Philonthus decorus*, density per m^2 in relation to winter moth (*Operophtera brumata*) pupae per m^2, where $h = 82.31 \pm 19.84$ (SE), $b = 88.70 \pm 58.10$ (SE) (data from Frank 1967).

and numerical responses.[7] For the numerical response, let us assume that the total population size of a generalist in the whole habitat is constant, buffered by its broad diet, but that in any one generation, t, only some of them, G_t, actually attack the insect species in question, depending on how abundant these hosts are. We assume this number is determined by the generalists 'switching' to feed preferentially on the host type that is currently the most abundant. What scant field evidence that exists suggests that this numerical response can be a relatively simple and immediate function of the host's density (Fig. 3.8):

$$G_t = h\left[1 - \exp\left(-\frac{N_t}{b}\right)\right]. \tag{3.4}$$

Here the constant h is the saturation number of generalists (i.e. the size of the pool of generalists available in the whole habitat), and the constant b determines where the inflexion of the curve occurs on the host density axis (Kowalski 1976; Southwood and Comins 1976). We now assume that each generalist has a type II functional response and that their distribution of attacks is aggregated and given by the negative binomial model. Host survival therefore becomes:

$$g(N_t, G_t) = \left[1 + \frac{a_g G_t}{k(1 + aT_h N_t)}\right]^{-k} \tag{3.5}$$

where a_g is the attack rate, T_h is the handling time (to give a maximum attack rate of $1/T_h$) and k is the degree of aggregation from the negative binomial distribution.

The net density dependence arising from this model is the result of two conflicting components: direct density dependence from the generalists' numerical response in eqn (3.4) and inverse density dependence from the type II functional response in eqn (3.5). The density dependent relationship therefore rises monotonically if the functional response is more-or-less linear over the range of prey densities, but becomes increasingly dome-shaped as the predators are increasingly limited by their maximum attack rate (Hassell and May 1986). Figures 3.7(a), (b) show the model fitted to examples from the field; the lack of any 'hump' in the data suggests that the functional responses tend to be roughly linear over the observed prey densities.

The host population dynamics can now be simply represented by:

$$N_{t+1} = \lambda N_t g(N_t, G_t) \tag{3.6}$$

and are illustrated in Fig. 3.9(a). Depending on the parameter values, there are a number of different possibilities:

1. There may be no equilibrium at all if the generalists are too inefficient (line A).
2. However, if they are efficient enough (particularly if a_g and h are high, and b and T_h are low) in relation to the host rate of increase (λ), there can be a locally stable host equilibrium (marked 'S' on lines B and C). The host population level at which this occurs depends in a predictable way on the various parameters of the model (low equilibria are promoted by high $a_g h$, and k, and low b and T_h, all enhancing predation, and low host rates of increase, λ).
3. If the density dependence is dome-shaped, and thus only operates over a range of host densities, there is a threshold host density (N_T) above which the host population 'escapes' from the generalist (line B). In this case we would expect additional density-dependent factors to come eventually into play once host densities reached high enough levels. If this density dependence is severe enough, maybe arising from scramble competition or an epidemic disease (May 1985), the higher equilibrium will be locally unstable and the populations will 'crash' back to low levels at which the generalists may once again regulate the population. Figure 3.9(b) gives such an example from the analysis of Clark's psyllid data (Clark 1963*a,b*, 1964*a,b*) by Southwood and Comins (1976).

Patterns of episodic, and short-lived, outbreaks consistent with Fig. 3.9(b) are characteristic of the dynamics of several insect herbivores (e.g. Fig. 1.1(b)) and are particularly well known from the gregarious larvae of forest defoliating sawflies and Macrolepidoptera (Hanski 1987). Conflicting mechanisms have been proposed to account for such dynamics. Rhoades (1985) has proposed a plant-defence explanation in which 'outbreak' herbivores are opportunistic species which tend to exploit their foodplants excessively and are thus

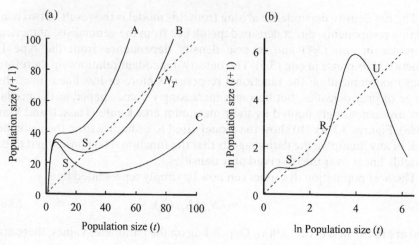

Fig. 3.9 Map of prey populations in successive generations, t and $t + 1$ driven by density dependence from generalist predators. (a) The three lines, A to C, are discussed in the text and are obtained from eqn (3.6) with survival from predation, g, defined in eqn (3.5) and parameter values as follows: $a = 0.15$, $h = 50$, $b = 30$, $\lambda = 20$ and $T_h = 0.2$ (A), 0.12 (B), 0.06 (C). Points marked S where lines B and C intersect the 45° line denote locally stable equilibria. The point marked N_T on line B denotes the threshold host density above which the host population 'escapes' from the generalist. (b) From a model of the population dynamics of the eucalyptus psyllid (*Cardiaspina albitextura*) in which S again indicates a locally stable equilibrium maintained by the natural enemies, R is the release point above which the population escapes from natural enemy regulation and U is an unstable equilibrium caused by overcompensating competition for resources. (From Southwood and Comins 1976).

much affected by variations in food quality. However, the more popular view is that the lower equilibrium is maintained by generalist natural enemies (usually predators), from which the herbivores occasionally 'escape' before food limitation and/or other kinds of natural enemies cause their rapid decline to the levels at which the generalists can again maintain their numbers (e.g. Holling 1973; Southwood and Comins 1976; Ludwig *et al.* 1978; Hassell and May 1986; Hanski 1987). Supporting evidence for this is mainly of two types. First, there are several examples of generalists causing density-dependent mortality, particularly of forest Lepidoptera pupae (Fig. 3.7), and secondly, and more specifically, there are examples suggesting that the density depend-ence operates over only a range of relatively low host densities; for example, in the spruce budworm, *Choristoneura fumiferana* (Holling 1973; Ludwig *et al.* 1978), the green rice leafhopper (Sasaba *et al.* 1973; Kiritani 1977) and in the reduviid bug, *Rhodnius prolixus*, which is a vector of Chagas disease (Rabinovich 1984).

3.3.4 Generalists and specialists together

In any natural community, insect herbivores are almost certain to be attacked by both generalist and specialist natural enemies. This leads to a particular combination of parasitism and host density dependence, the interplay of which can potentially lead to dynamics that are much more complex than those illustrated in Fig. 3.5 and Fig. 3.9. This section reviews a classic long-term study in which the relative roles of generalist predators and specialist parasitoids have been examined, and demonstrates the range of dynamics that can be obtained from a more broadly based model.

The winter moth

One of the classic long-term studies in insect ecology, and one in which generalist predators are thought to have had a major impact on population dynamics, is that of Varley and Gradwell on the winter moth, *Operophtera brumata*, in Wytham Wood, Oxford, England (Varley and Gradwell 1960, 1963*a*, 1968; Varley *et al.* 1973). Their census of the winter moth populations began in 1950 and continued uninterrupted until 1968. From these data they developed their 'key-factor' method for analysing sequences of life-table data from organisms with discrete generations (Varley and Gradwell 1960). Key factor analysis has since been widely used in life-table studies on insects (e.g. Klomp 1966; Harcourt 1971; McNeill 1973; Nakamura and Ohgushi 1981), fish (Elliott 1984), birds (Blank *et al.* 1967; Krebs 1970; Southern 1970; Watson 1971), mammals (Sinclair 1973) and even plants (Silvertown 1982), and their original method has been developed and refined by several people (Podoler and Rogers 1975; Manly 1990; Sibly and Smith 1998). Varley and Gradwell were also able to parameterise a simple model for their winter moth population, and were the first to attempt to do this from a long-term insect study.

The winter moth is a univoltine species. The adults emerge from the soil in November and December. The wingless females climb a nearby tree trunk, mate with the winged males and then continue upwards to oviposit in crevices, or under lichen, in the crown of the tree. Egg-hatch typically occurs in early April and the newly eclosed larvae then seek to enter the opening leaf buds. The fifth and final instar is found from early to late May depending on the weather conditions, and is followed by a prepupal stage which spin to the ground on silken threads and pupate in the surface layers of the soil where they remain until emerging again as adults in the following autumn. Varley and Gradwell studied the winter moth population in a stand of oak trees and, for their routine census data, restricted their samples to five particular oak trees. Sampling was carried out by intercepting a fraction of the population at two stages of the life cycle: adult females as they ascended the trees (using 'lobster-pot' style, tree trunk traps), and prepupae as they descended to the ground (in two 0.5-m^2 caterpillar trays placed under each tree). The prepupae were then dissected to determine the incidence of parasitism and disease. In this way they were able to recognise, in each generation, a number of mortalities or 'disappearances' between stages. These can be conveniently divided into the five survival categories below, which were quantified in each year of the study.

'Winter disappearance' (s_1)

This is made up of the combined losses of winter moth occurring between Varley and Gradwell's estimates of egg density and their subsequent counts of total prepupae descending to the ground to pupate. Although this includes a wide range of factors affecting natality and mortality (e.g. predation of adults by shrews and spiders, variations in fecundity per adult, predation of eggs by beetles, harvestmen and others, bird predation of larvae and larval parasitism), the major component was certainly the huge loss immediately following egg-hatch when the majority of newly emerged larvae fail to get established within an opening leaf bud. It is a highly variable mortality from year to year (mean $s_1 = 0.11$; range $= 0.03$ to 0.58), making it the 'key factor' primarily responsible for determining population fluctuations. It depends largely on the cumulative day-length and weather conditions during winter and spring which affect the degree of synchrony between egg-hatch and bud-burst: heaviest mortalities happen when egg-hatch occurs well before bud-burst, and vice versa. In the absence of any means of predicting the s_1-values, Varley and Gradwell included it in their models either as a constant equal to the mean or as the actual observed values (see below).

Survival from Cyzenis albicans (s_2)

Varley and Gradwell were very interested in the role of parasitoids in the winter moth population dynamics, and paid particular attention to larval parasitism by a tachinid fly, *Cyzenis albicans* (Hassell 1968, 1969*a,b*, 1980*a*; Varley *et al.* 1973). *Cyzenis* at Wytham Wood is almost entirely a specialist parasitoid of winter moth larvae. The adult flies emerge from the soil at the end of March or early April, before their hosts (the fifth-instar winter moth larvae) are available. They then feed at flowers, on sap fluxes and honeydew so that by the time that the winter moths are in their fifth instar, the female *Cyzenis* have matured their full complement of 1000 to 2000 eggs which are crammed into an expanded oviduct ready for laying. But, instead of ovipositing directly in or on their hosts, the females lay their eggs on the foliage upon which the winter moth feeds, and in particular around areas of leaf damage (Hassell 1968). The eggs hatch if ingested by a feeding winter moth,[8] and parasitism occurs by a first-instar *Cyzenis* larva burrowing through the midgut wall and then into the winter moth's salivary gland. Here it remains until reaching its second instar after the winter moth has pupated in the soil. It then develops rapidly through the second and third instars, killing the winter moth and forming a puparium within the winter moth cocoon. The *Cyzenis* pupa then remains in the soil until the following year when the adults of the next generation emerge.

By dissecting samples of winter moth prepupae descending to the ground to pupate, the frequency distribution of parasitism by *Cyzenis* was determined in each year; and over the whole period was found to be much better described by the negative binomial than the Poisson distribution (see Fig. 2.4). Winter moth survival from parasitism by *Cyzenis* (s_2) is therefore given by the zero term of the distribution:

$$s_2 = \frac{N}{N - N_a} = \left(1 + \frac{aP}{k}\right)^{-k} \tag{3.7}$$

where N is the density of fifth-instar winter moth larvae, N_a is the density of larvae parasitised by *Cyzenis*, k is the clumping parameter from the negative binomial distribution (mean $k = 1.09$) and a is the parasitoid searching efficiency (mean $a = 0.18 \text{ m}^2$ per generation).[9]

Survival from other parasitoids (s_3) and from a microsporidian (s_4)

There are other, relatively minor, factors affecting the survival of the fifth-instar larvae descending to the ground to pupate. In particular, a few other parasitoid species and a microsporidian disease (*Plistiphora operophterae*) attack the larvae. The magnitude of these varied little from year to year and their mean values (mean $s_3 = 0.97$ and mean $s_4 = 0.93$) are thus used in the winter moth model below.

Survival from natural enemies in the soil (s_5 and s_c)

Pupal mortality of winter moths in the soil is a large mortality caused mainly by predation by carabid and staphylinid beetles and small mammals (Frank 1967; Buckner 1969), and by the ichneumonid parasitoid, *Cratichneumon culex* (Varley *et al.* 1973) (but excluding parasitism by *Cyzenis* which is scored as a larval mortality). It is density-dependent as shown in Fig. 3.7(c), and the overall trend is well described by the simple relationship:

$$s_5 = 0.60 N_t^{-0.31} \tag{3.8}$$

where N_t is the density of winter moth prepupae per m^{-2} descending to the ground to pupate.[10]

The same predators that eat winter moth pupae also attack other pupae in soil. The immature *Cyzenis*, both the larvae within parasitised winter moth cocoons and subsequently the puparia, are therefore vulnerable to these predators and have been shown to suffer a similar density-dependent mortality (Fig. 3.2), which has been described by Hassell (1980a):

$$s_c = 0.16 N_a^{-0.37} \tag{3.9}$$

where N_a is the density of parasitised winter moth as given by eqn (3.7).[11] As we shall see below, this is an important feature of the dynamics of the interaction.

The winter moth model

Varley and Gradwell modelled the winter moth larval population as a discrete generation host–parasitoid interaction, akin to model (3.3) but with additional survival terms added (Varley and Gradwell 1963*a*, 1968) as described above:

$$N_{t+1} = \lambda N_t s_1 s_2 s_3 s_4 s_5$$
$$P_{t+1} = s_c N_t (1 - s_2). \tag{3.10}$$

The resulting dynamics are illustrated in Fig. 3.10, from which Varley and Gradwell drew a number of conclusions. Not surprisingly, the correspondence between observed and calculated populations since 1950 is very close when the observed s_1-values are included in the model. This just emphasises the extent to which this is the key factor driving the population fluctuations. Making s_1 constant removes all these fluctuations from the model and reveals the underlying stable equilibrium. That this is due to the density-dependent winter moth pupal mortality, s_5, rather than the parasitoid, *Cyzenis*, is made clear by completely eliminating *Cyzenis* from the model: the increase in winter moth densities is barely detectable. In short, the winter moth at Wytham Wood is regulated by the density-dependent pupal mortality; it fluctuates primarily due to the overwintering mortality; and it is little affected by the parasitoid *Cyzenis*. The reason that *Cyzenis* is so ineffective is because of the very heavy

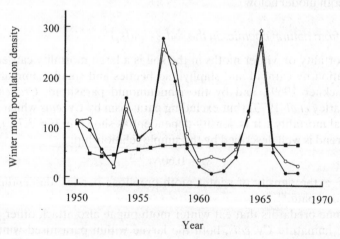

Fig. 3.10 The observed fluctuations in winter moth larval/prepupal population density (per m² of the canopy area) (solid circles) at Wytham Wood compared with predicted populations using eqn (3.10) parameterised as described in the text. Hollow circles: predicted populations using observed s_1-values. Solid squares: predicted populations using mean s_1-values of 0.11. (From Hassell 1980*a*.)

density-dependent mortality that it suffers in the soil (eqn (3.9)). This provides a graphic example of the importance of gathering life-table information on natural enemies as well as their hosts or prey. Without the survival of *Cyzenis* having been quantified, its complete lack of effect on the winter moth population would not have been predicted.

It is interesting to explore how different the interaction would be if *Cyzenis* were freed from this heavy mortality while in the soil (i.e. making $s_c = 1$). The predicted winter moth equilibrium would then be reduced from $N^* = 80.3$ to $N^* = 11.2$, with a tendency for damped oscillations (Fig. 3.11(a)). The stability arises in part from the density dependence acting on the winter moth pupae and in part from the aggregated distribution of parasitism by *Cyzenis* ($k = 1.09$). This can be demonstrated by increasing the value of k for *Cyzenis* (moving towards random search). Instead of being stable, the winter moth population now cycles considerably (Fig. 3.11(b)), and at even higher values of k is unable to maintain itself and goes extinct.

The winter moth has also been studied extensively in Canada where it was accidentally introduced in the early 1930s (Embree 1966, 1971). In 1954 a biological-control programme was initiated in Nova Scotia in which six parasitoid species imported from Europe were released (Graham 1958), and of these, two species, the tachinid fly (*Cyzenis*) and an ichneumonid parasitoid (*Agrypon flaveolatum*), became established. As levels of parasitism built up there was a dramatic decline in the winter moth populations. This has been attributed to the high levels of parasitism achieved by the *Cyzenis* population

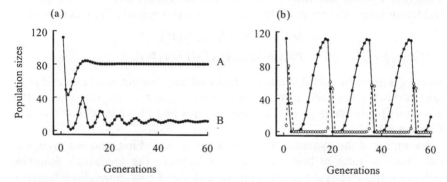

Fig. 3.11 Simulated numbers of winter moths per m² canopy area using eqn (3.10). (a) Line A: as in Fig. 3.10 (solid squares) in which *Cyzenis* suffers a strong density dependent mortality in the soil (see eqn (3.9)). Line B: as for line A except for the absence of any mortality imposed on *Cyzenis* ($s_c = 1$). The parasitoids are no longer ineffectual and considerably reduce winter moth numbers (from $N^* = 80.3$ to $N^* = 11.2$) and also combine with the density-dependent soil mortality to promote population stability. (b) Simulated winter moth (solid circles) and *Cyzenis* (hollow circles) populations as for line B, except that the parasitoids now search at random and are therefore destabilising. The populations persist only because of the density dependence acting on the winter moth in the soil.

operating without the constraints of heavy density-dependent predation (Embree 1965, 1966), and therefore acting as an effective biological-control agent in first reducing and then maintaining the winter moth populations at low levels (Hassell 1980a). Roland (1986, 1988, 1989, 1994), however, has re-examined these data in the light of his studies on the winter moth and *Cyzenis* in British Columbia and come to a rather different conclusion. It appears that earlier studies in Canada overestimated the importance of *Cyzenis* and under-estimated the impact of the mortality in the soil. Roland concludes that, while parasitoids were important to the initial decline of winter moth populations in Nova Scotia, it is the generalist predators in the soil that are maintaining the populations at their current low levels.

A general framework

The winter moth example in the previous section is a special case of generalist and specialist natural enemies acting together. Here we examine a more general framework showing how interacting generalists and specialists can lead to unexpectedly complex dynamics.

Let us consider the case of a host population in which the specialist parasit-oids in each generation only attack those hosts that have survived attack from the generalists. This could occur if the generalists attack an earlier host stage than the specialists or, equivalently, if they search for the same host stage but always consume the specialist should the same host individual be attacked by both (see Hassell and May (1986) for further details and for a comparable treatment where the order of these mortalities is reversed). We can now write:

$$N_{t+1} = \lambda N_t g(N_t, G_t) f(P_t)$$
$$P_{t+1} = N_t g(N_t, G_t)[1 - f(P_t)]$$

$$(3.11)$$

where g is given by eqn (3.5) and f is the usual negative binomial expression for host survival from parasitism (an important implicit assumption here is that any clumping in the degree of aggregation in g and f is uncorrelated (see also Chapter 6.3)).

Examples of the dynamics are shown in Fig. 3.12. First, and not unexpect-edly, for low rates of host increase, λ, relative to the specialist's searching efficiency, the specialist can never invade and only a host–generalist interaction is possible (broken lines). Quite simply, there is too little host production to support a viable population of specialists, especially with the generalist acting first in the host's life cycle. The same effect occurs if the generalists are too effective (due to high values of $a_g h$ from eqns (3.4) and (3.5)) relative to the efficiency of the parasitoids. Otherwise, the specialist can readily invade to give a three-species system, especially if the generalists concentrate their attacks in an aggregated way on the host population (i.e. k is small), so leaving a larger pool of hosts for the specialist to exploit. Much more surprising is the range of alternative states shown in Fig. 3.12 that arise when, despite highly

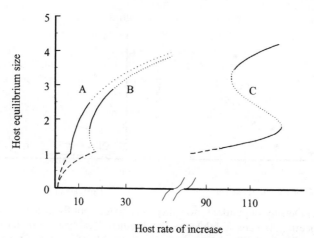

Host rate of increase

Fig. 3.12 Three examples of the dynamical properties of model (3.11) showing the dependence of the host equilibrium and its stability on the host rate of increase (λ). For each curve the broken line indicates a two-species, host–generalist interaction. Elsewhere, there is a three-species system with hosts, generalists and specialists which can either be locally stable (solid lines) or locally unstable (dotted lines). The searching efficiencies of the specialist and generalist equal one and the distribution of attacks is random ($k \to \infty$). Other parameter values from eqn (3.4) as follows: (curve A) $h = 20$ and $b = 10$; (curve B) $h = 30$ and $b = 10$; (curve C) $h = 5$ and $b = 0.4$. (After Hassell and May 1986.)

efficient generalists, the specialist is still able to invade because of a sufficiently high host rate of increase. For example, C shows two alternative stable interactions (solid lines) separated by a locally unstable region (dotted lines).

Such alternative stable states are an intriguing possibility when strong, direct density dependence from generalists is combined with the time-lagged density dependence from specialists. Although not yet firmly identified from any real system, these dynamics do appear in a parameterised model based on a cabbage root fly population (*Delia radicum*) attacked by a generalist staphylinid beetle, *Aleochara bilineata*, and a specialist cynipid wasp, *Trybliographa rapae* (see Jones *et al.* 1993 for details). Figure 3.13(a) shows the generalist's density dependence which, in the absence of the parasitoid *Trybliographa*, leads to stable *D. radicum* populations (Fig. 3.13(b), line A). But with *T. rapae* also present, three-species alternative stable states exist for a wide range of possible values for the host rate of increase (Fig. 3.13, lines C and D).

3.4 Summary

Most basic host–parasitoid models assume that only a single adult parasitoid can emerge from a host, irrespective of how many parasitoid eggs may be laid

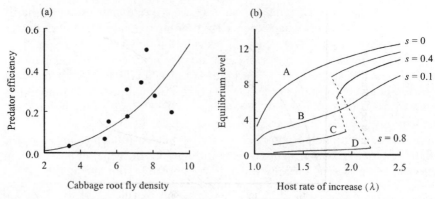

Fig. 3.13 (a) Density-dependent mortality of cabbage root fly puparia caused by the staphylinid beetle, *Aleochara bilineata*. (b) Predicted cabbage root fly equilibrium densities in relation to the flies' rate of increase (λ), with *Aleochara* as the only natural enemy acting (line A) and with both *Aleochara* and the parasitoid *Trybliographa rapae* present at three different levels of parasitoid survival (s) as shown (lines B, C and D). The broken lines are drawn to link the alternative equilibrium states. (After Jones *et al.* 1993, in which full details are given.)

in that host, implying strong, compensating density dependence acting on the parasitoid population. This is very important to the stability properties. For example, the condition for stability that $k < 1$ in May's negative binomial model depends on this assumption. If competition between parasitoid larvae within a host is stronger, stability is possible for weaker levels of aggregation of parasitoid attacks ($k > 1$); but if competition is weaker, then stability requires very much stronger aggregation ($k \ll 1$). Finally, in the limit that there is no larval competition (i.e. a complete lack of any density dependence), stability is no longer possible for *any* level of aggregation.

Density dependence can also act on a parasitoid population in other ways. For example, there may be density-dependent shifts in the parasitoid sex ratio or the parasitised hosts may suffer a density-dependent mortality. Anything affecting the parasitoids in this way will contribute to the stability of the interaction.

The assumption of a constant host rate of increase is usually made so that the dynamical effects of parasitism can be more easily discerned. Most host populations, however, will suffer density dependence at some stage in their life cycles, often stemming either from competition for resources or the action of generalist natural enemies. In the absence of parasitoids, the effects of this density dependence are relatively straightforward. For example, in the case of an insect herbivore in a patchy habitat with density-dependent survival of larvae within each patch, stability is enhanced by stronger within-patch density dependence and higher degrees of patch-to-patch aggregation.

This picture is complicated when parasitism and host density dependence are combined. The dynamics now depend on how close is the host equilibrium to the parasitoid-free carrying capacity. If close, the dynamics are primarily driven by the host density dependence (bottom-up regulation). Otherwise, the host–parasitoid dynamics increasingly prevail (top-down regulation).

Generalist natural enemies can cause inverse or direct density-dependent mortality amongst their prey. The latter is especially likely if their numbers remain relatively constant in relation to any one of their prey types, and if they tend to 'switch' between prey depending on their relative abundance. This switching produces a numerical response which, when combined with an appropriate functional response, gives the overall density-dependent response of the generalists. Within an insect–generalist model, the generalists, if efficient enough, can maintain a locally stable host population, but above a threshold density, the hosts may 'escape' from this regulation until other processes curb their increase. When specialist parasitoids are added to the interaction, the dynamics become much more complicated. One or the other of the natural enemies may be unable to persist (depending on their relative searching efficiencies). Alternatively, if all three species coexist, then there may be a variety of alternative stable states. In short, the combination of different kinds of natural enemies introduces a wider range of dynamics than expected from a blending of the separate pairwise components. Chapter 6 develops this theme and examines the properties of more complex host–parasitoid communities.

Notes

1. Some parasitoids lay 'microtype' eggs on the foliage of the host's food plant. Parasitism then occurs when an egg is ingested by a host.
2. An exhaustive field study in which *no* evidence of density-dependent sex ratio shifts has been found is that of Reeve and Murdoch (1986) on the California red scale and its parasitoid *Aphytis melinus* (see Chapter 5).
3. These terms parallel to some extent 'exploitation' and 'interference' competition. 'Scramble' and 'contest', however, are used more to describe the dynamical outcome of competition while 'exploitation' and 'interference' refer more to the behavioural mechanisms involved.
4. Alternatively, this condition can be expressed as $aK > 1$, which makes close contact with the criterion for invasion ($R_0 > 1$) in basic models of host–pathogen interactions (e.g. Anderson and May 1993).
5. Recently, a full treatment of the global properties of this model with the inclusion of a type II functional response has been given by Kaitala *et al.* (1999) showing how complicated the dynamics can be when one moves from a local analysis to a global one. The complexities include multiple attractors, basins of attraction with fractal properties, chaotic-like transients and chaotic attractors.
6. The density dependence in Fig. 3.7(c) is of this kind. Winter moth larvae are eaten by generalist predators (mainly carabids and staphylinids) irrespective of whether or not they contain larvae of the parasitic fly, *Cyzenis albicans*.
7. Solomon (1949) defined numerical responses as the relationship between the numbers of predators and prey density.

8. The eggs will hatch in the midgut of a few other species (e.g. *Cosmia trapezina*) but are only known to survive in the larvae of winter moth and its relative, *Operophtera fagata*.

9. Because Varley and Gradwell also collected some life-table data on the *Cyzenis* population, including the survival of parasitoid progeny and the density of emerging adult *Cyzenis* each generation, they were able to use eqn (3.7) to estimate parasitoid searching efficiencies (see Varley *et al.* (1973) for full details of their study).

10. Varley and Gradwell (1968, 1971) modelled *Cratichneumon* separately, leading to a three-species, host–parasitoid interaction. Subsuming it within the total soil mortality greatly simplifies their model without affecting the major conclusions.

11. The same exponents in eqns (3.9) and (3.10) reflect the common cause of the winter moth and *Cyzenis* mortalities. The higher level of mortality in eqn (3.10) occurs because the *Cyzenis* puparia are in the soil for much longer than the winter moth pupae.

4

Heterogeneity in host–parasitoid interactions

4.1 Introduction

With the exception of the negative binomial model of parasitism (Griffiths 1969b; May 1978), the host–parasitoid models considered in the last two chapters viewed populations as homogeneous entities in which roaming natural enemies encountered hosts at random, like molecules colliding in an ideal gas (Thompson 1924; Nicholson and Bailey 1935; Watt 1959; Hassell and Varley 1969). This tradition of assuming that all host individuals are equally susceptible to parasitism by identical parasitoids is an obvious starting point for developing any general theory, but it is highly simplistic. Within any natural population, individuals of varying phenotype and phenology will almost inevitably be unevenly distributed within habitats that are highly structured and variable. It is therefore inconceivable that individuals in real populations will all experience the same probability of survival and reproduction throughout their lives.

No study has had a greater influence in publicising how heterogeneity can affect population dynamics than C. B. Huffaker's classic experiments with predatory and prey mites feeding on oranges (Huffaker 1958; Huffaker et al. 1963). In simple environments where relatively few oranges were interspersed with rubber balls, prey and predator populations rapidly became extinct (Fig. 4.1(a)). But in a much more divided environment with 120 oranges, each with 5% of their surface area exposed, and with dispersal impeded by Vaseline barriers but promoted by air currents and launching sticks, the populations persisted and showed pronounced predator–prey cycles (Fig. 4.1(b)).

While Huffaker's experiments tell us little about the mechanisms by which spatial patchiness promotes persistence, they have been the inspiration for the development of many models for spatially structured predator–prey systems (e.g. Hilborn 1975; Caswell 1978; Hastings 1978; Crowley 1979; Nisbet and Gurney 1982). A comparable host–parasitoid experiment is shown in Fig. 4.2; the interaction is unstable in a more-or-less homogeneous environment, but persists in a relatively stable interaction when the beans, which are the host resource, are confined within small patches. In both these examples, increased partitioning of the environment into discrete patches reduced the chances of

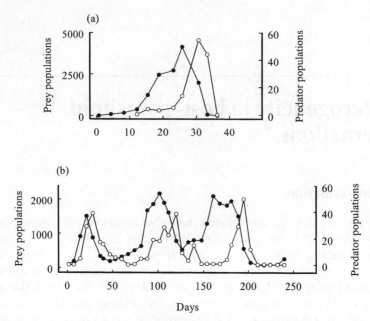

Fig. 4.1 Predator–prey interactions between the predatory mite, *Typhlodromus occidentalis*, (hollow circles) and its prey mite, *Eotetranychus sexmaculatus*, (solid circles) feeding on oranges in a laboratory system. (a) A single oscillation followed by predator and prey extinction in a simple system, and (b) sustained oscillations from a much more complex system. (After Huffaker 1958; Huffaker *et al.* 1963.)

extinction and allowed the populations to persist at levels well below the host or prey's carrying capacity.

This chapter explores the effects of spatial and other kinds of heterogeneity on the dynamics of host–parasitoid systems. In doing this, we are straightaway confronted with the problem of the vastly different spatial scales over which individuals and populations interact. For example, groups of *individuals* may occur within discrete patches (often foodplants) within their habitat; these are collectively gathered into *local populations*, which, when interconnected by restricted dispersal, form *metapopulations*, which in turn can be interlinked at a *regional* scale by the very occasional interchange of dispersing individuals. The distinction between local populations and metapopulations is an important one. Local populations are characterised by considerable mixing of their individuals in each generation, so that no part of the population shows significant autonomous dynamics over time. Within a metapopulation, however, dispersal between its constituent parts—the local populations—is much more restricted and this allows the local populations to develop asynchronous dynamics. This chapter only considers scales at the local

(a)

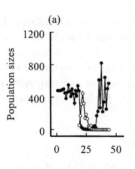

(b)

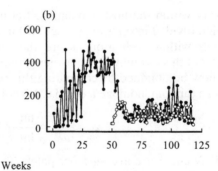

Weeks

Fig. 4.2 Population dynamics of the bruchid beetle, *Callosobruchus chinensis*, (solid circles) feeding on black-eyed beans and its pteromalid parasitoid, *Anisopteromalus calandrae*, (hollow circles) in a laboratory system (Hassell and May 1988). (a) A non-patchy system with 50 beans uniformly distributed on the arena floor. The parasitoids are introduced in week 19 and become extinct in week 32 allowing the hosts to increase until checked by resources. (b) A patchy system with 50 beans each in an individual container with restricted access to both hosts and parasitoids.

population level and below; metapopulation dynamics are discussed in Chapter 7.

4.2 What is heterogeneity?

'Heterogeneity' has become something of a catchall term in ecology. Its use in this chapter, however, will be quite specific. Following Chesson and Murdoch (1986), it is defined in terms of the variation in risk of parasitism between different individuals in the host population. In this terminology, therefore, a habitat is only 'heterogeneous' in so far as it leads to an 'aggregation of risk' of parasitism between host individuals. The importance of this measure lies in the way that it can quantify the stabilising effect of an aggregated distribution of parasitism amongst the host individuals in a population. Chesson and Murdoch first noted that the instantaneous probability of a host being parasitised from the Nicholson–Bailey model is given by (see eqn (2.5)):

$$aP = \ln\left(\frac{N}{S}\right) \tag{4.1}$$

where N, P and S are respectively, the hosts, searching parasitoids and survivors from parasitism in generation t^1. Because the Nicholson–Bailey model assumes both homogeneous populations and a constant searching efficiency, a, all host individuals are exposed to the same number, P, of parasitoids and therefore share the same risk of parasitism. However, if a were to vary between hosts (or groups of hosts within patches), or some hosts were exposed to more searching parasitoids than others, this risk of parasitism would vary within the host population. Let us consider, for example, a host population divided into j groups, each within a distinct patch, where the host individuals within each group share the same risk of parasitism, $a_j P_j = \ln(N_j/S_j)$, but this risk may vary between groups. The *relative risk of parasitism*, ρ, can now be obtained by comparing this risk *within* a group with the risk assuming a random, homogeneous (i.e. Nicholson–Bailey) interaction:

$$\rho = \frac{a_j P_j}{a\bar{P}} = \frac{\ln(N_j/S_j)}{\ln(\bar{N}_j/\bar{S}_j)} \tag{4.2}$$

where $\bar{P}$, $\bar{N}$ and $\bar{S}$ are averages per patch. The distribution of ρ-values thus represents the variability in the relative risk of parasitism between hosts. For example, the Nicholson–Bailey model is recovered if ρ is uniformly distributed per patch, while the negative binomial model of May (1978) is obtained when ρ is gamma-distributed (see p. 75). It is this variability that is the key to understanding the stabilising effects of heterogeneity in the interactions described in this chapter.

One of the most obvious ways in which heterogeneity of risk of parasitism can arise is in a patchy environment where the level of parasitism varies between patches; much of this chapter concentrates on this scenario. But it can also arise in quite different ways. For example, host and parasitoid life cycles may not be properly synchronised so that some hosts are less at risk from parasitism, or escape completely, due to the phenological mismatch of life cycles that do not coincide (see p. 89). Or, there may be phenotypic variation between host individuals such that some hosts are able to reduce their risk of parasitism by virtue of their physiology or behaviour. This is considered below (p. 92). Heterogeneity of risk is thus a pervasive feature of natural interactions.

A very clear exposition of how this heterogeneity helps to stabilise host–parasitoid interactions of the form of model (2.1) is given by Taylor (1993b). Briefly, the *within*-generation effects of heterogeneity of risk depend solely on the function $f(\cdot)$ describing host survival from parasitism. This translates into a stabilising effect *between* generations by the dependence of $f(\cdot)$ on parasitoid density per generation, P_t. In short, aggregation of risk leads to pseudo-interference, which has already been described in Chapter 2. Taylor also describes a number of situations when this mechanism for stability breaks down: for instance, when surviving parasitoid progeny per parasitised host is proportional to the number of times the host has been encountered (Taylor 1988a, and see Section 3.2), when the parasitoid population is uncoupled from that of the host and when the interactions are in continuous time rather than discrete generations (see Chapter 5).

4.3 Spatial patterns of parasitism

Let us commence with the same patchy environment described in Chapter 3, made up of discrete patches of foodplants upon which an insect herbivore species feeds. The herbivore has discrete generations and is attacked by a specialist parasitoid species whose life cycle is synchronised with that of the hosts. Searching adult parasitoids therefore coincide temporally with the susceptible host stage. On emergence, the adult hosts disperse from their natal patch and having mated, the females then move amongst the patches ovipositing. The resulting immature hosts are confined to their respective patches until the next generation of adults emerges. Like the adult hosts, the emerging adult parasitoids mix thoroughly prior to foraging for their hosts within the patches. The scale of the habitat is therefore small enough, and the mixing thorough enough, for these to be local rather than metapopulations.

The end result of this behaviour is that the total populations of N_t hosts and P_t adult female parasitoids are distributed amongst the n patches such that within any one patch there are P_i parasitoids searching for N_i hosts. Suppose first that the parasitoids, rather than aggregating in the more profitable patches, divide themselves evenly, and that within any patch they have a linear functional response, a constant searching efficiency and encounter hosts at random. Per cent parasitism is now the same in each patch in any one generation, and the Nicholson–Bailey model for the whole population is recovered exactly. The spatial distribution of hosts between patches is irrelevant in this case since all hosts, irrespective of the patch that they are in, suffer the same risk of parasitism.

Such a uniform risk of parasitism across patches, whilst convenient mathematically, is countered by a wealth of evidence from both laboratory and field studies. For instance, the three laboratory examples in Fig. 4.3 and Fig. 4.4 are clear cases where the searching parasitoids spend more time, and therefore tend to congregate, in the patches of high host density. In the two cases in Fig. 4.3 the resulting patterns of parasitism are density-dependent (Figs 4.3(b), (d)), but in the third case, in Fig. 4.4(b), it is just the opposite—parasitism is inversely density-dependent despite the parasitoids aggregating in the high, host density patches. This difference stems from the different functional responses of the three species. The two parasitoids in Fig. 4.3, *Trybliographa rapae* and *Venturia canescens*, have relatively short handling times giving relatively high, maximum attack rates per parasitoid (i.e. high upper asymptotes of their functional responses). Having more parasitoids per patch thus translates into a higher percentage of parasitism, and hence the density-dependent patterns in Figs 4.3(b) and (d). The egg parasitoid in Fig. 4.4, *Trichogramma pretiosum*, however, has a much longer handling time and hence a lower maximum attack rate. Individual parasitoids are therefore restricted in their ability to exploit high host density patches, and inverse density dependence arises even though the searching parasitoids aggregate in response to host density. Examples illustrating this mechanism are shown in Fig. 4.5.

Data of such detail, combining information on the distribution of searching parasitoids and the resulting patterns of parasitoids, are very difficult to collect

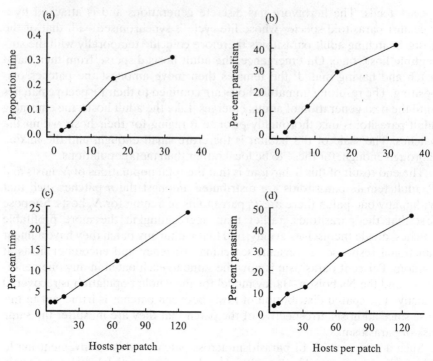

Fig. 4.3 Average patterns of time allocation (a, c) and parasitism (b, d) per patch in relation to host density per patch from two laboratory systems. In both cases, the density-dependent pattern of parasitism reflects the distribution of searching effort by the parasitoids on the different patches. (a, b) Ten females of the cynipid parasitoid, *Trybliographa rapae*, parasitising larvae of the cabbage root fly (*Delia radicum*) at different densities within discs of swede (Jones and Hassell 1988). (c, d) Eight females of the icheumonid parasitoid, *Venturia canescens*, parasitising larvae of the flour moth, *Ephestia cautella*, at different densities within circular containers on the cage floor (Hassell 1982).

from the field (but see Waage 1983; Driessen and Hemerik 1991). On the other hand, it is relatively easy just to quantify the patterns of parasitism without bothering about the distribution of searching adults: hosts need only be sampled from a range of patches, taken back to the laboratory and the incidence of parasitism then determined, usually by rearing out the next generation or by dissection. The data obtained in this way fall into three distinct categories, showing direct, inverse or density-independent relationships with host density per patch (Fig. 4.6). Of the 201 different examples listed in the reviews of Lessells (1985), Stiling (1987), Walde and Murdoch (1988) and Hassell and Pacala (1990), 59 show direct density-dependent patterns of parasitism, 53 show inverse patterns and 89 show parasitism uncorrelated with host density per patch.

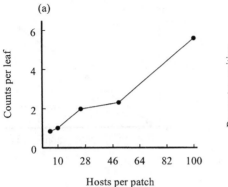
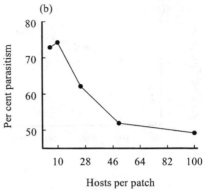

Fig. 4.4 Average patterns of time allocation (a) and parasitism (b) per patch in a laboratory system with 16 females of the trichogrammatid parasitoid, *Trichogramma pretiosum*, parasitising eggs of the stored product moth, *Plodia interpunctella*, stuck on brussels sprout leaves at different densities. In contrast to Fig. 4.3 parasitism is inversely density-dependent despite the tendency for the parasitoids to aggregate on leaves with the highest host densities (Hassell 1982).

Using the terminology of Chesson and Murdoch (1986), the positive and negative density-dependent patterns correspond to their 'pure regression' models in which the distribution of parasitoids is a deterministic function of local host density, and the density-independent patterns correspond to their 'pure error' models in which the density of searching parasitoids in each patch is a random variable independent of local host density. Thus, in a regression of local parasitoids on local host density, covariance between the densities of parasitoids and hosts is represented by the regression function, while variation in the densities of parasitoids that is independent of host density is accounted for by the regression error term. Pacala *et al.* (1990) and Hassell *et al.* (1991b) proposed a similar dichotomy in which the direct and inverse patterns arise from '*host density-dependent heterogeneity*' (HDD) and the density-independent patterns from '*host density-independent heterogeneity*' (HDI). This terminology is more broadly descriptive of the patterns in Fig. 4.6 without putting the emphasis on the nature of the regression equation.

There have been different views about how these patterns arise. Hassell (1982) accounted for both direct and inverse patterns in terms of the maximum attack rate of the functional response, as illustrated in Fig. 4.5. Similarly, but taking a more evolutionary perspective, Lessells (1985) showed that parasitoids maximising the rate of finding healthy hosts would cause direct density-dependent parasitism unless parasitoid fitness was time- or egg-limited, in which case dome-shaped or inverse patterns would result. More complex patterns would occur if the parasitoids were able to assess host density as they search within a patch (Iwasa *et al.* 1981; Lessells 1985). As well as the details of foraging behaviour and functional response, spatial scale can be important in

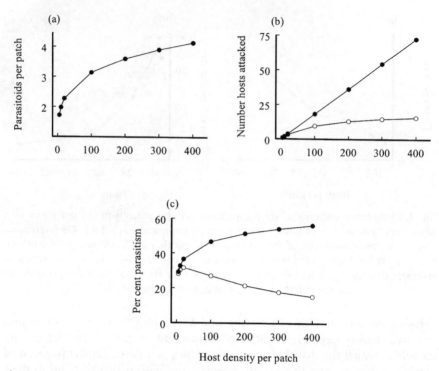

Fig. 4.5 Numerical examples showing how the combination of aggregating parasitoids and different kinds of functional response can lead to both density-dependent and inverse density-dependent spatial patterns of parasitism. (a) An arbitrary aggregated distribution of searching parasitoids. (b) Two kinds of functional response defining the attack rate per parasitoid within a patch. Solid circles from a linear functional response with $a = 0.2$ and $T_h = 0$; hollow circles from a type II response with $a = 0.2$ and $T_h = 0.05$. (c) The resulting overall per cent parasitism per patch with solid and hollow circles corresponding to those in (b). Notice that the type II response more than compensates for the aggregation of the parasitoids to cause inverse density dependence.

determining these patterns. Bernstein *et al.* (1991) thought that parasitoids would not be good at tracking differences in host abundance at either very large or very fine scales, making HDI patterns more likely in such cases. They would be much better at tracking host densities, however, at intermediate spatial scales at which one should therefore expect a preponderance of HDD patterns (which again should be positive unless a strongly limiting functional response outweighs the aggregation of searching adults). But this does not concur with Walde and Murdoch's (1988) findings from their analysis of 75 examples of spatially distributed parasitism. HDD patterns occurred at a wide range of spatial scales but in those studies performed on small scales in which 'small samples were taken from points scattered across the habitat without regard to their spatial distribution within the larger collective unit', a clear

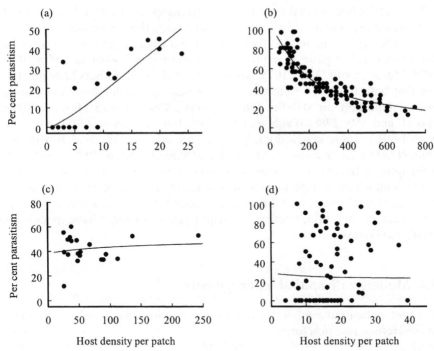

Fig. 4.6 Examples of field studies showing different spatial patterns of parasitism. (a) Density-dependent parasitism of the larvae of cabbage root fly (*Delia radicum*) by the cynipid parasitoid, *Trybliographa rapae* (Jones and Hassell 1988). (b) Inverse density-dependent parasitism of gypsy moth (*Lymantria dispar*) eggs by the encyrtid parasitoid, *Ooencyrtus kuwanai* (Brown and Cameron 1979). (c) Density-independent parasitism of the olive scale (*Parlatoria oleae*) by the aphelinid parasitoid, *Coccophagoides utilis* (Murdoch *et al.* 1984). (d) Density-independent parasitism of the gall midge, *Rhopalomyia californica*, by the torymid parasitoid, *Torymus baccaridis* (Ehler 1987). See text for a discussion of the fitted lines. (From Pacala and Hassell 1991.)

majority fell into the inverse category. Where samples were collected within larger units, however, the majority of HDD patterns found were positively related to patch density. They concluded that the inverse patterns resulted primarily from limiting functional responses, and that the direct patterns within the larger units were caused by reproductive build-up in areas of high host density of parasitoids that did not disperse much from generation to generation. This therefore invokes elements of a metapopulation.

In short, categorising the conditions for obtaining HDD and HDI patterns is likely to be difficult. But we can say that the form of the functional responses in relation to the aggregation of the adult parasitoids is likely to be important in determining whether the spatial density dependence is direct or inverse, and that at larger metapopulation scales differential reproduction and dispersal will become increasingly important.

Much of the interest in these different patterns has focused on their potential importance to population dynamics, for which quite different claims have been made. The earlier theoretical literature tended to emphasise the primacy of density-dependent patterns in promoting stability (e.g. Hassell and May 1973, 1974; Murdoch and Oaten 1975; Beddington *et al.* 1978). Others have pointed out that both the inverse patterns *and* the density-independent ones can be just as important for stability (Hassell 1984a; Chesson and Murdoch 1986; Hassell and May 1988; Walde and Murdoch 1988; Pacala *et al.* 1990; Hassell *et al.* 1991b). These conclusions, however, are all based on models with discrete generations. In contrast, Murdoch and Stewart-Oaten (1989) concluded, from quite different models in continuous time, that density-independent patterns may have no effect on stability, while the density-dependent ones can be destabilising! The rest of this chapter attempts to reconcile this apparently contradictory literature, starting first with a review of some basic models for HDD and HDI parasitism.

4.4 Models with spatial heterogeneity

The host–parasitoid models described in the previous chapters have all taken the discrete-generation form:

$$N_{t+1} = \lambda N_t f(N_t, P_t)$$
$$P_{t+1} = cN_t[1 - f(N_t, P_t)]. \tag{4.3}$$

This framework can also be applied to interactions in a patchy environment as long as the function $f(N_t, P_t)$ represents the *average*, across all patches, of the fraction of hosts escaping parasitism. In any explicit representation of patchiness, therefore, $f(N_t, P_t)$ must depend on both survival from parasitism *within* patches as well as the spatial distributions of hosts and parasitoids *between* patches. We commence with the very simple scenario of a habitat with n patches, within each of which parasitism is random and determined by a type II functional response. The distribution of hosts and parasitoids from patch to patch can follow any distribution defined by α_i and β_i, the fractions of total hosts and total searching parasitoids, respectively, in the ith patch. Thus, f is given by:

$$f(N_t, P_t) = \sum_{i=1}^{n}\left[\alpha_i \exp\left(-\frac{a\beta_i P_t}{1 + aT_h\alpha_i N_t}\right)\right] \tag{4.4}$$

where a is the searching efficiency per patch and T_h is the handling time (Hassell and May 1973). Notice again that the Nicholson–Bailey model is recovered in the limit that the parasitoids have a linear functional response and treat all patches equally (P_t/n parasitoids in each patch) so giving a uniform risk of parasitism across all patches. Once the parasitoids aggregate in some patches over others, however, stability is enhanced and the general criterion for local stability, assuming $T_h = 0$, is given by:

$$\lambda \sum_{i=1}^{n} [\alpha_i(a\beta_i P^*) \exp(-a\beta_i P^*)] < \frac{\lambda - 1}{\lambda}. \tag{4.5}$$

Specific examples of this model, where the distribution of parasitoids leads either to spatial density dependence (HDD) or density independence (HDI), are described in the next sections. They fall into two broad categories: (1) 'fixed aggregation' models where there is one episode of redistribution at the beginning of the parasitism period, after which the distributions are 'fossilised' until the next generation; and (2) 'continuous redistribution' models where the parasitoids change their distribution within the period of parasitism as they move around from patch to patch.

4.4.1 Fixed aggregation models

Density-dependent aggregation (HDD)

The direct and inverse spatial patterns of parasitism shown in Fig. 4.6 may reflect the actual distributions of the searching parasitoids (if the functional responses are more-or-less linear), or they may be strongly influenced by both the parasitoid distribution *and* the functional responses if the latter are strongly non-linear (Fig. 4.5). For simplicity, we will assume linear responses so that the patterns of parasitism mirror the distribution of adult parasitoids (Hassell 1984a). We also assume that the density-dependent aggregation of parasitoids is given by the simple power relationship:

$$\beta_i = w\alpha_i^{\mu} \tag{4.6}$$

where μ is an index of parasitoid aggregation and w is a normalisation constant such that the β_i values sum to unity (Hassell and May 1973). The index, μ, can describe a wide range of parasitoid distribution patterns: the searching parasitoids will be evenly distributed across patches if $\mu = 0$, and will increasingly tend to aggregate in patches of high host density as μ rises until, in the limit $\mu \to \infty$, they will all congregate in the single patch of highest host density leaving the remainder as complete refuges (see p. 88). Following MacArthur and Levins (1964) and MacArthur (1968), the parasitoids are perfectly 'fine-grained' in their distribution when $\mu = 1$ (i.e. equal fractions of hosts and parasitoids in each patch) and increasingly 'coarse-grained' as μ increases or decreases (Soberon 1986). Lastly, if $\mu < 0$ the parasitoid distribution is reversed, with their local abundance now inversely correlated with host density per patch. The way that these different patterns can promote stability is shown by the following example. We assume that the hosts are aggregated according to a negative binomial distribution, giving:

$$f(N_t, P_t) = N_t \sum_{j=0}^{\infty} [p(j)j\exp(-aP(j))] \tag{4.7}$$

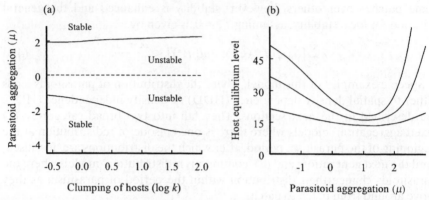

Fig. 4.7 The effects of parasitoid aggregation on dynamics. (a) Stability boundaries between the degree of aggregation of parasitoids, μ, and the amount of host clumping, k, (assuming a negative binomial distribution of hosts per patch) from model (4.3) with survival from parasitism, $f(\cdot)$, given by (4.7), total number of patches $n = 30$, searching efficiency $a = 1$ and host rate of increase $\lambda = 2$. (b) Examples of the host equilibrium level varying with the degree of parasitoid aggregation, μ, for three different numbers of patches (n) from model (4.3) with $f(\cdot)$ given by (4.4) and (4.6), $a = 0.5, \lambda = 2$ and $n = 11$ (top curve), $n = 7$ (middle curve) and $n = 5$ (lower curve). (After Hassell 1984a, which gives full details.)

where $p(j)$ is the probability of having j hosts in a patch from the negative binomial distribution and $P(j)$ are the number of parasitoids in a patch with j hosts (defined by eqn (4.6)). An example of the stability boundaries from this model is given in Fig. 4.7(a). Notice that sufficiently strong direct *or* inverse density-dependent distributions of parasitoids can stabilise the interactions. The interactions are locally unstable if there is insufficient parasitoid aggregation, or if the host rate of increase is above some threshold level,[2] in which case a range of interesting global dynamics occurs (Rohani *et al.* 1994b).

Finally, Fig. 4.7(b) shows how these different degrees of parasitoid aggregation affect equilibrium levels. The parasitoids have the greatest effect in reducing host equilibria when their distribution most closely tracks that of the hosts (i.e. $\mu = 1$). Increasing parasitoid aggregation (positive or negative) leads to higher host abundances, simply because the parasitoids are constrained by a fixed aggregation strategy; once distributed they are confined to their respective patches, irrespective of the degree of host exploitation.

Density-independent aggregation (HDI)

There has been much more emphasis on understanding, modelling and measuring HDD than HDI heterogeneity, partly because of the pivotal role it has been assumed to have in promoting population stability, and partly because behavioural ecologists wished to make the link between foraging behaviour and population dynamics (Godfray 1994). It is now clear, however, that HDI heterogeneity can be at least as important a process in promoting population persistence.

Let us consider a specific example where all the heterogeneity in parasitism is independent of the spatial distribution of hosts (HDI), as in Fig. 4.6(d). We will assume that this is achieved by the distribution of searching parasitoids being *aggregated* from patch to patch *independently* of the host density per patch (and in this particular case following a gamma distribution). The host distribution is therefore irrelevant. Finally, we assume that within any one patch the parasitoids exploit hosts at random with a constant per capita searching efficiency (linear functional response). The fraction of hosts escaping parasitism is now given by:

$$f(P_t) = N_t \int_0^\infty g(\varepsilon)\exp(-aP_t\varepsilon)d\varepsilon \qquad (4.8)$$

where $g(\varepsilon)$ is the gamma probability density function for parasitoids per patch with unit mean and variance $1/\alpha$ (α is a positive constant governing the shape of the density function) and a is the usual per capita searching efficiency of the parasitoid. In each patch, therefore, host survival by $P_t\varepsilon$ randomly searching parasitoids is given by the zero term of a Poisson distribution with mean $aP_t\varepsilon$ (Murdoch *et al.* 1984a; Chesson and Murdoch 1986). Very much the same derivation would apply were the parasitoids uniformly distributed across patches but their searching efficiency, a, varied according to a gamma distribution (Bailey *et al.* 1962). The stability properties of the model are straightforward: the interaction will be stabilised by the HDI heterogeneity as long as there is sufficient variance in the distribution of parasitoids or, more specifically, if $\alpha < 1$.

Apart from being a model of pure HDI parasitism, eqn (4.8) is also interesting for the way that it reduces exactly to May's negative binomial model (see Chapter 2, p. 19). What was previously a purely phenomenological index of aggregation, *k,* is now explicitly related to the degree of density-independent aggregation of the searching parasitoids (α) (Chesson and Murdoch 1986; Pacala *et al.* 1990; Hassell *et al.* 1991b). The stability criterion, *k* < 1 from the negative binomial model, therefore corresponds exactly in this spatially explicit model to $\alpha < 1$. Similar stabilising properties have also been demonstrated from related models with density-independent parasitism by Reeve *et al.* (1989), further emphasising that density-independent patterns of parasitism are potentially important as a stabilising mechanism. This is explored in more detail in Section 4.5.

4.4.2 Continuous redistribution models

Optimal foraging

The parasitoids in the previous sections had an inflexible aggregation strategy. At the start of each generation they distribute themselves amongst the patches where they then remain confined, irrespective of how heavily the hosts become exploited and irrespective of whether or not there are greener pastures elsewhere. The parasitoids in this section are much more adaptive in being able continually to seek out the best patches currently available. The large body of literature on optimal patch-use stems from the classic paper by Charnov (1976) in which a forager leaves a patch when the instantaneous rate of gaining fitness (via food, reproductive sites, etc.) is reduced to the maximum rate achievable in the environment as a whole (e.g. Stephens and Krebs 1986; Krebs and Kacelnik 1991; Krivan 1996, 1997).

Turning specifically to parasitoids, for any given spatial distribution of hosts, and depending on behavioural and life history constraints, there is an optimal distribution of parasitoid searching time that maximises the rate of parasitism for the individual searching parasitoids. Cook and Hubbard (1977) and Hubbard and Cook (1978) developed a model for a population of optimally searching parasitoids that were able unfailingly to optimise their distribution amongst patches to maximise the rate of discovering unparasitised hosts. Comins and Hassell (1979) produced exactly the same results, but from a different model in which each individual parasitoid moves amongst patches (with zero travelling time), perfectly allocating its searching time given a particular host distribution and constant searching efficiency and handling time (see also Royama 1971; Lessells 1985). Consequently, at the start of the searching period, each parasitoid is found in the patch with the highest host density. As soon as the rate of encountering healthy hosts in this patch is reduced to the level of encounter that could be achieved in the second most profitable patch, the parasitoids divide their search between patches 1 and 2 (but not equally if the handling time is greater than zero since more time will be 'wasted' in patch 1 in handling already parasitised hosts). So the process continues, with the set of exploited patches broadening until, if there were sufficient parasitoids with enough time, all patches would be reduced to the same level of profitability. This is illustrated by the example in Fig. 4.8(a). Therefore, unlike the fixed aggregation model above, the relative distribution of parasitoids changes during the searching season, being most aggregated at the start and becoming less so as time progresses.

The dynamical effects of this continuous redistribution and evening-out of the numbers of parasitoids per patch is shown in Fig. 4.8(b) in terms of the overall searching efficiency in relation to the numbers of parasitoids searching.[3] The horizontal, broken line is for randomly searching parasitoids; their probability of finding a host (healthy or parasitised) is the same in all patches and throughout the searching season. The searching efficiency therefore remains constant irrespective of the total parasitoid searching effort, *PT*. Curve B is for parasitoids with a fixed aggregation strategy. At low to intermediate values of *PT* the parasitoids are more efficient than their random counterparts because of their aggregation in the patches of high host density (assumed $\mu = 1$). But at higher values of *PT* their inability to move away from

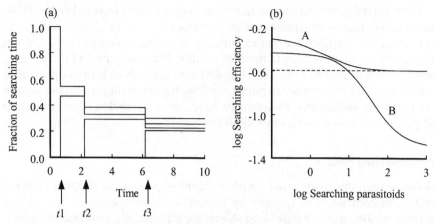

Fig. 4.8 Results from an optimal foraging model for a single parasitoid in a 4-patch system with a host distribution of 10, 6, 3 and 1 hosts per patch. (a) Temporal sequence of parasitoid searching time on the four patches with searching efficiency per patch, a_p = 1, and handling time, T_h = 0.05. Until time t_1 all parasitoid time is spent on the patch with 10 hosts. The set then broadens to include the patch with 6 hosts, and so on until after t_3 all four patches are being exploited. (b) Apparent interference relationships showing the decline in searching efficiency (log a) in relation to the log density of searching parasitoids (or more strictly the product of parasitoid density and searching time, PT) from two different models in which the host distribution and searching efficiency are as in (a) but T_h = 0. Each is compared with the Nicholson–Bailey result (broken line) where there is no change in searching efficiency as parasitoid density changes. Curve A: Optimal foraging as described in the text. Curve B: Parasitoids show fixed aggregation described by eqn (4.6) where μ = 1. (After Comins and Hassell 1979.)

these patches once well exploited greatly reduces their efficiency to the point that they fare even worse than the random searchers. Finally, the optimal foragers (Curve A) are the most efficient at low PT when they all congregate in the most profitable patches. But as PT increases and they tend to reduce all patches to the same profitability, their efficiency comes closer and closer to that of the random parasitoids. The contribution of all these patterns to stability depends on the negative slope of these lines evaluated at the PT equilibrium (see Chapter 2, p. 27). Thus if the parasitoid equilibrium is very high or if the period of search, T, is very long, the optimally foraging parasitoids will contribute very little to stability, much as if they were searching randomly. This is in contrast to the fixed aggregation parasitoids which are always stabilising. However, at lower values of PT, optimal foraging leads to aggregated distributions of parasitoids and HDD parasitism, which in turn can be a powerful mechanism for stability. In short, the stabilising effects of optimal foraging depend on how much the aggregation of risk early in the searching season has been reduced by the end of the season (Hassell and Pacala 1990).

More broadly, we can conclude that the contribution to stability by 'HDD parasitoids' that redistribute within generations will tend to decrease as parasitoid density rises and as the duration of the searching period increases. This is discussed more fully in the next section. However, most of the patterns of parasitism discussed so far (e.g. Fig. 4.6) have come from host samples taken *after* the period of exposure to parasitoids; so it seems unlikely that parasitoids are generally sufficiently abundant or have enough searching time to reduce all patches to similar levels of profitability.

An interesting debate

We have seen that aggregated spatial patterns of parasitism, whether HDD or HDI, can contribute to stability by generating heterogeneity in the risk of parasitism. But this has only been shown for interactions with discrete generations in which, with the exception of the optimal foraging model above, the dispersing stages of hosts and parasitoids distribute themselves amongst the patches at the start of each generation and then remain in these positions throughout the season. Different model structures, however, can lead to quite different conclusions. For example, Murdoch and Stewart-Oaten (1989) examined a model whose limiting case has hosts and parasitoids interacting in continuous time with perfectly overlapping generations and infinite dispersal rates. Both HDI and HDD aggregation are modelled in terms of the variance in the parasitoid distribution and covariance in the distributions of hosts and parasitoids, respectively. Within this framework, HDI heterogeneity has *no effect* on stability at all, while HDD heterogeneity either contributes little to stability or, more likely, is *destabilising*.

The stark contrast between these two sets of conclusions has made for an interesting debate. For example, Ives (1992a) developed a continuous time model that is similar to Murdoch and Stewart-Oaten's but is a different limiting case in which the susceptible host stage is infinitely short. In effect this means that there is no continuous host dispersal and local parasitism depends only on the local density of searching parasitoids at that moment. As in the discrete-generation models, stability now hinges once again on the extent of variation in the risk of parasitism. Godfray and Pacala (1992) took issue with the conclusion that density-independent aggregation has only a neutral effect, arguing that this stems from the unrealistic limiting case of infinite host dispersal rates. Murdoch *et al.* (1992a) responded by developing a continuous-time metapopulation model in which further intriguing properties emerge; for example, density-dependent aggregation can be stabilising or destabilising, with stronger aggregation leading to *less* stability by decreasing the asynchrony between the host patches (metapopulation host–parasitoid models are discussed in Chapter 7).

In an attempt to clarify how much the presence, or absence, of within-generation redistribution is at the core of these very different predictions,

Rohani *et al.* (1994*a*) adapted a model from Godfray *et al.* (1994) which retains discrete generations, but allows more-or-less continuous redistribution of the parasitoids during the generation period (see p. 91). In this respect the model is in the same spirit as the optimal foraging model above, but differs in having an explicit age-structure for the populations (see Chapter 5) and in allowing both HDD and HDI parasitism between the patches. The model is divided into separate within- and between-generation periods. Between generations, the assumptions are the same as for the models above: the adult hosts emerging from the previous generation move freely amongst the patches laying eggs such that the fraction of the total host eggs in patch *i* is given by α_i. Similarly, the emerging adult female parasitoids colonise patches such that the fraction in the *i*th patch at the *start* of the 'season' is given by $\beta_i = w\alpha_i^\mu$, as in eqn (4.6), where μ defines the level of parasitoid aggregation at the start of the searching period (for HDI aggregation, $\mu = 0$). The within-generation period is formulated as an age-structured, continuous-time model (see Chapter 5). Instead of the initial distributions now being fossilised throughout the within-generation period of parasitism, the parasitoids continually leave patches at a rate *M*. For simplicity, travel time between patches is assumed to be zero, so that all parasitoids leaving a patch instantaneously enter another patch with a probability that is assumed constant if aggregation is HDI, but varies according to some aggregation index if parasitism is HDD. The juvenile hosts, on the other hand, do not redisperse within a generation; as in the majority of juvenile insect species, they are relatively immobile and remain in their natal patches.

Full details of the model are given in Rohani *et al.* (1994*a*). The most straightforward case is provided by HDI aggregation. Its stabilising effect is not influenced at all by within-generation parasitoid movement; the net movements into and out of patches are equal and thus preserve the aggregated distribution of parasitoids across patches as the within-season interaction proceeds.[4] This is not the case, however, with HDD aggregation, where within-generation redistribution certainly does alter the effect on stability (Fig. 4.9). Just as with the optimal foraging parasitoids in the previous section, the differences between patches in the risk of parasitism at the start of the season (due to parasitoids aggregating in the high host-density patches) is progressively reduced by within-generation movement. In other words, as time progresses, the remaining healthy hosts are more evenly spread out in the population and there is less variation in the risk of parasitism.

This analysis therefore does not support Murdoch and Stewart-Oaten's (1989) conclusion that HDI parasitism has no effect in promoting stability; rather we find that the stabilising effect of HDI aggregation is not influenced at all by within-season movement. It does give some support, however, to their view that the impact of HDD aggregation has been exaggerated in the past. As in the optimal foraging model in the previous section, it appears that HDD aggregation will only be stabilising if the period of search is short enough,

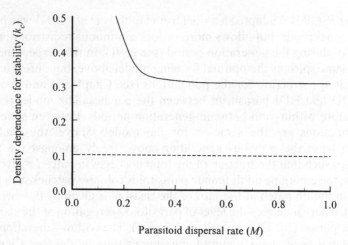

Parasitoid dispersal rate (*M*)

Fig. 4.9 The effects of increasing within-season parasitoid movement between patches on the amount of parasitoid density dependence required for stability (k_c) in a five-patch model in which the fraction of hosts per patch is 0.4, 0.3, 0.2, 0.05, 0.05 and the host larvae are susceptible for 10 time units (*w*) within the season. At the start of the season the parasitoids aggregate in the highest density patch following eqn (4.6) with $\mu = 5$. Thereafter during the season the parasitoids disperse at a rate, *M*, and dispersing individuals continue to aggregate with $\mu = 5$. Explicit density dependence acting on the parasitoid is introduced by defining the per capita risk of parasitism from:

$$k \ln\left(1 + \frac{aP(t)}{k}\right)$$

where *a* is the usual parasitoid attack rate and *k* is a measure of the dependence of parasitoid efficiency on parasitoid density (thus small values of *k* increase the stabilising effect—see Godfray and Hassell 1989; Godfray *et al.* 1994 for full details). The solid line in the figure shows the increased amount of parasitoid density dependence needed for stability as the amount of seasonal redistribution increases. This is to be compared with the situations: (1) where there is no parasitoid aggregation or redistribution ($\mu = M = 0$), in which case $k = 0.1$ is required for stability (thus $kw = 1$ as in May's (1978) negative binomial model), shown by the broken line in the figure; and (2) where there *is* aggregation ($\mu = 5$) but *no* redistribution, in which case almost no parasitoid density dependence is needed for stability. (After Rohani *et al.* 1994*a*.)

and/or the dispersal rate of parasitoids small enough, to preserve at the end of the season sufficient of the initial heterogeneity in the risk of parasitism that was established at the start of the searching period. Hence, parasitoids that tend to be confined in their search to a single patch or which make relatively few patch visits in their life are the best candidates to be influenced by density-dependent aggregation. The data on this are mixed. Of the 108 examples of HDD parasitism referred to in Section 4.3 above, most of them record parasitism after the searching period has finished. The levels of heterogeneity recorded are therefore good measures of their potential contribution in promoting stability (which is quantified in the next section). Other examples record the behaviour and distribution of the searching adult parasitoids. For example, the parasitoid, *Leptopilina clavipes*, studied by Driessen and Hemerik (1991; 1992), makes only a few patch visits during its searching lifetime. In this case, any HDD heterogeneity is likely to be largely conserved throughout the searching period. In contrast, Jones *et al.* (1996) studied host and parasitoid movement from a field study on tephritid flies and their parasitoids, and found, using mark–recapture techniques, that parasitoid movement was on a scale likely to prevent any tendency for HDD parasitism to be conserved.

Identifying qualitatively different patterns of parasitism and assessing their potential for population regulation are first steps in understanding the importance of this kind of heterogeneity for population dynamics. The next stage is to try and quantify this effect for particular data sets collected in the field.

4.5 A more unified approach

The previous sections have emphasised the potential importance of an aggregated risk of parasitism, irrespective of how this may arise. Despite the common currency, however, it remains difficult to evaluate the contribution that different patterns of parasitism may make to this heterogeneity and hence to the stability of natural host–parasitoid interactions. This section reviews some first steps in attempting to clarify the situation, first by outlining a general criterion that shows the contribution of HDD and HDI parasitism to stability, and second by applying this approach to data that are readily available in the field.

We start by turning back to the model for HDI aggregation described above (see p. 74), in which the parasitoids aggregate independently of local host density according to a gamma distribution with a unit mean density per patch and variance of $1/\alpha = 1/k$ (stable as long as $k < 1$), where α is a positive constant determining the shape of the density function. An alternative way of expressing this stability criterion makes use of the square of the coefficient of variation of searching parasitoids per patch ($CV^2 = \text{variance/mean}^2$), which in this case is simply defined by $1/k$. The stability condition $k < 1$ is therefore identical to the condition $CV^2 > 1$ (May 1978; Chesson and Murdoch 1986; Hassell and May 1988; Pacala *et al.* 1990); in other words, the interaction will be stable if the distribution of searching parasitoids per patch, measured as the square of the coefficient of variation (CV^2) is greater than one.[5]

Interestingly, this specific criterion approximates quite well to the requirements for stability across a broad spectrum of discrete-generation, host–parasitoid models (Pacala *et al.* 1990; Hassell *et al.* 1991b). But first, the way that the CV^2 is measured needs to be modified. The problem with a definition based just on the variance in the numbers of searching parasitoids per patch is the high degree of scale-dependence. For instance, consider the example in

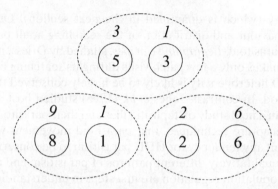

Fig. 4.10 Diagram illustrating different estimates of CV^2, depending on the spatial scale chosen. The small circles indicate patches with number of hosts as shown within. The parasitoids searching on each patch are shown above in italics. The CV^2 calculated from the distribution of parasitoids across the five patches is 0.76. Different definitions of patch boundaries leads to different values of CV^2. Thus CV^2 becomes 0.17 if calculated over three patches as shown by the broken lines. Finally, the CV^2 weighted for the different numbers of hosts per patch is 0.36. See text for further details.

Fig. 4.10 of six patches with 5, 3, 8, 1, 2 and 6 hosts per patch, on which forage 3, 1, 9, 1, 2 and 6 parasitoids, respectively. The CV^2 across the six patches is 0.76; but this value will change if the same population is grouped in a different way (e.g. $CV^2 = 0.17$ if the populations are grouped into the three larger patches shown). This difficulty can be avoided if the CV^2 is always weighted by the number of hosts in the different patches. Now there are 25 hosts in the six patches and the CV^2 is estimated across these 25 entities.[6] So the 5 hosts in the first patch all experience 3 parasitoids, the 3 hosts in the next patch experience 1 parasitoid, and so on giving a weighted CV^2 of 0.36. In this way, by using information on the number of searching parasitoids exposed to each individual host, we are returning closer to the recurrent theme emphasising the variation in risk between the different individual hosts.

A number of discrete-generation models have been analysed to see how well the stability criterion, $CV^2 > 1$, applies (Pacala *et al.* 1990; Hassell *et al.* 1991b), three of which span the spectrum of HDI and HDD patterns of parasitism.

1. *Model I (pure HDI)*
This is the HDI example above, in which $CV^2 > 1$ exactly represents the condition for stability.

2. *Model II (pure HDD)*
From one extreme of no correlation between the host and searching parasitoid distributions, Model II goes to the other extreme of a perfect correlation between the two. Specifically, the hosts are aggregated following a gamma

distribution, and the parasitoids deterministically track this patch-to-patch variation in local host density using the aggregation function in eqn (4.6). Thus, there is direct density-dependent aggregation in patches of high host density if $\mu > 0$ and inverse density-dependent aggregation if $\mu < 0$. The $CV^2 > 1$ rule is now only approximately true, but the approximation is good if the hosts are highly aggregated and have a low net rate of increase.

3. *Model III*

Models 1 and II make specific assumptions about the host and parasitoid distributions and represent specific end-points on the continuum between HDI and HDD models. In general, there will be both density-dependent and density-independent components to the spatial distribution of parasitoids. In much the same spirit as Reeve *et al.* (1989) who explored a model combining both HDD and HDI patterns of parasitoid aggregation, Model III allows the distribution of relative host numbers in patches to take any form whatsoever—as long as it does not change with total host density, and defines the parasitoid distribution by an arbitrary function that combines HDD and HDI aggregation. Stability is now approximated by the slightly more complex expression, $CV^2 - 1/p^* > 1$ where p^* is the average number of parasitoids that visit a patch at equilibrium, calculated with respect to a randomly chosen host. The $1/p^*$ term is important since it represents the component of CV^2 that is caused solely by purely random (Poisson) variation in parasitoid abundance among patches. Such variation does not contribute to aggregation of risk and therefore does not contribute to stability. For example, if the average number of parasitoid visits per patch is only one, then $1/p^* = 1$ and the stability criterion becomes $CV^2 > 2$. But if, on average, there are many parasitoid visits per patch, then the $CV^2 > 1$ rule once again applies reasonably well. In short, the HDI component of heterogeneity only contributes to stability in so far as it includes non-random aggregation of parasitoids between patches.

4.5.1 Applications to field data

A useful feature of the $CV^2 > 1$ rule is that it is readily decomposed into its constituent parts of HDD and HDI heterogeneity, both of which can be quantified directly from empirical data collected in the field. The basis of this is the assumption that the distribution of searching parasitoids is given by the same power function as for the HDD Model II, to which is added the gamma-distributed residual, ε, as in the HDI Model I:

$$p = cP_t \left(\frac{n}{N_t} \right)^{\mu} \varepsilon \tag{4.9}$$

where n and p are the local host and parasitoid densities, respectively. Thus, the magnitude of any HDD aggregation is given by the value of μ, and HDI aggregation by the magnitude of the variance of ε. Full details of the analysis are given in Pacala and Hassell (1991). Their key conclusion is that the $CV^2 > 1$ rule may be approximated as:

$$CV^2 \approx C_I C_D - 1 \qquad (4.10)$$

where C_I is the HDI component given by $C_I = 1 + \sigma^2$ in which σ^2 is the variance of ε, and CD is the HDD component given by $CD = 1 + V^2\mu^2$ in which V is the weighted coefficient of variation of the host density per patch. The $CV^2 > 1$ rule therefore applies if $C_I C_D > 2$, which can arise from HDI alone if $C_I > 2$, HDD alone if $CD > 2$, or some combination of the two.

Applying the $CV^2 > 1$ rule to field data is relatively straightforward in the few cases where there are data on the actual distribution of searching parasitoids in relation to host density per patch. Driessen and Hemerick (1991) provide a good example. They were able to observe directly the foraging of *Leptopilina clavipes*, an eucoilid parasitoid of *Drosophila* larvae in stinkhorn fungi, and from these data estimated the overall CV^2 directly from the mean and variance of the distribution of parasitoid observations. The CV^2 values fell between 0.23 and 0.53, indicating relatively little stabilising heterogeneity. Interestingly, they found that the distribution of parasitism, as measured by the negative binomial distribution, was highly aggregated, suggesting that factors other than the distribution of searching adults were contributing to the overall aggregation of risk. Reeve *et al.* (1994b) came to a similar conclusion from their study of the saltmarsh planthopper, *Prokelisia marginata*, and its egg parasitoid, *Anagrus delicatus*. They estimated CV^2 from the distribution of parasitism (see below) and found that $CV^2 > 1$ for 24 out of 28 sample dates, indicating strongly stabilising heterogeneity.[7] But the distribution of adult *Anagrus* was not sufficiently aggregated to account for this, once again suggesting that the aggregation of risk arose largely from other factors.

Estimating the stabilising effects of HDD and HDI heterogeneity from the distribution of searching adult parasitoids suffers from two major drawbacks. First, the data are hard to collect in the field, and second, as the examples above show, there are clearly factors other than the distribution of adults determining heterogeneity in the risk of parasitism. It is possible, however, to estimate CV^2 directly from the distribution of parasitism rather than from the distribution of adult parasitoids. This has two main advantages: (1) the data are far easier to collect and are widely available in the literature; and (2) the estimates will encompass other factors that contribute to aggregation of risk. The procedure for doing this is based on *inferring* the distribution of searching parasitoids from the pattern of parasitism, assuming the two are linked by a linear functional response (Pacala and Hassell 1991). In other words, it works back to what the CV^2 of adult parasitoids per patch would have had to have been if their distribution were the sole determinant of the observed levels of parasitism and if their functional responses were linear. In practice, it involves estimating the key parameters in eqn (4.10) from data on levels of parasitism and host density per patch: μ and σ^2 from maximum likelihood estimates, and

V^2 directly from the data on the distribution of hosts. C_I and C_D are thus obtained and thence the value of CV^2 (full details are given in Pacala and Hassell 1991).

The values of C_I and C_D, and thence CV^2, have been estimated for the four examples in Fig. 4.6. In (a) there is a clear direct density-dependent pattern of parasitism. The estimated CV^2 of 1.16 indicates that such a level of heterogeneity in parasitism, were it typical from generation to generation, should be just sufficient to stabilise an interaction of the form of eqns (4.3) without the need for any additional stabilising mechanisms. The relatively large value for C_D ($= 2.06$) and the small value for C_I ($= 1.05$) indicate that virtually all of this stabilising heterogeneity comes from the density-dependent parasitism (HDD), which in turn depends upon *both* the relatively high value of the parasitoid aggregation index ($\mu = 1.37$) *and* the highly aggregated spatial distribution of the host ($V^2 = 0.66$). With insufficient host clumping, no amount of HDD parasitism would be enough to make $CV^2 > 1$. The importance of sufficient host clumping is illustrated by the example in (b). In this case there is pronounced inverse spatial density dependence ($\mu = -0.95$), but the relatively weak host clumping ($V^2 = 0.25$) leads to low values of C_D ($= 1.29$). Because the HDI heterogeneity in this example is also weak ($C_I = 1.11$), the spatial pattern of parasitism (again, if repeated from generation to generation) would contribute little to stabilising heterogeneity ($CV^2 = 0.37$). The remaining two data sets, (c) and (d) show no evidence of density-dependent parasitism. In (c) there is also no appreciable effect of HDI heterogeneity. But in (d) the HDI variation is very pronounced ($C_I = 8.25$), producing marked stabilising heterogeneity ($CV^2 = 7.33$) (note that this does not depend at all on the host's spatial distribution). Thus, although appearing erratic, the data in (d) actually contain more evidence of factors that could stabilise dynamics than do any of the previous examples.

In an analysis of 65 such data sets (Hassell and Pacala 1990; Pacala *et al.* 1990), $CV^2 < 1$ in 47 of them, thus indicating rather weak levels of stabilising heterogeneity. In the remaining 18 cases, $CV^2 > 1$, indicating that heterogeneity, if consistently at that level, ought to be sufficient to stabilise the populations. Interestingly, in the great majority of these, HDD was insignificant and HDI heterogeneity alone was sufficient to make $CV^2 > 1$. Only in four examples was HDD the dominant kind of heterogeneity and sufficient for $CV^2 > 1$. In short, and contrary to what is still a popular view, this suggests that while both density-dependent and density-independent patterns of parasitism can contribute to the stability of host–parasitoid interactions, it is the density-independent ones that are the more important.

Most of the examples above come from field studies with little, if any, spatial and temporal replication. In particular, the crucially important information on the *typical* levels of heterogeneity over a period of time is usually lacking. In the few cases where CV^2 has been estimated over a period of time, it has been found to be quite variable. For example, Redfern *et al.* (1992) found $CV^2 > 1$

in four out of six generations for parasitism of the tephritid fly *Urophora stylata* but in only one out of five generations for parasitism of another tephritid *Terralia serratulae*. Jones *et al.* (1993), in a study of the natural enemies of the cabbage root fly (*Delia radicum*), found $CV^2 > 1$ in only one year out of nine for parasitism by the cynipid parasitoid *Trybliographa rapae*, and in three years out of nine for parasitism by the staphylinid beetle *Aleochara bilineata*.

4.5.2 Some caveats

The $CV^2 > 1$ rule provides a rough indication of the amount of stabilising heterogeneity in discrete-generation, coupled host–parasitoid interactions. It also has the merit of being readily divided into HDD and HDI heterogeneity, both of which can be estimated from available field data on the distribution of parasitism. But several factors will affect the applicability and validity of the rule, and an excellent review of these is given by Taylor (1993b). The main caveats, in no particular order, are listed below.

- Most of the available information comes from single-site studies carried out within a single generation. The extrapolation to population dynamics is made by assuming that this 'snapshot' of estimates of CV^2 and host distribution are typical of the patterns from generation to generation, which will clearly not always be the case. Most examples where $CV^2 > 1$ have been found should thus really be interpreted as levels of heterogeneity of risk which, if typical over a period of time, would be sufficient to stabilise the host–parasitoid interaction without the need for any other stabilising mechanisms. The rule is thus a means of evaluating the *potential* contribution of observed levels of heterogeneity.
- Most natural host–parasitoid systems involve a number of competing host and parasitoid species, and the extent to which the rule may be applied to total parasitism of a host population by several species of parasitoids is unknown.
- Estimates of CV^2 from field data on parasitism involve extrapolating from parasitism per patch to the density of searching adult parasitoids, assuming a linear functional response and hence a close correspondence between the distribution of searching parasitoids and parasitism. Ives (1992b), however, has emphasised that the relationship between searching parasitoids and parasitism can be much weaker if the parasitoids have type II functional responses. More heterogeneity will then be needed to overcome the destabilising effect of the type II functional response.
- As well as temporal variability in spatial heterogeneity, the $CV^2 > 1$ rule will also be confounded by the presence of other important density-dependent processes acting upon the host population, due to resource limitation, generalist predators, pathogens and so on. In general, and not surprisingly, less heterogeneity of risk will be needed if these other processes contribute to the stability of the host population.

- The $CV^2 > 1$ rule assumes that there is random exploitation of hosts within patches; in other words, all the heterogeneity is assumed to arise from patch to patch. However, if parasitism is also aggregated *within* patches, this in itself is stabilising and thus reduces the critical amount of between-patch heterogeneity needed for stability (Comins *et al.* 1992). Suppose, for example, that host survival within a patch is described by the zero term of the negative binomial distribution. The CV^2 rule is now more complex and becomes $(1 + CV^2)(1 + 1/k) > 2$. Thus, no *between*-patch heterogeneity at all is needed if *within*-patch parasitism is sufficiently non-random (i.e. $k < 1$). But, as k increases, the stabilising effect of the within-patch aggregation lessens and the value of CV^2 needed for stability increases towards one (for example, for $k = 2$ we have the '$CV^2 > 0.33$ rule').

- Finally, the $CV^2 > 1$ rule has been formulated using discrete-generation models with no redistribution. We have seen that the importance of HDD heterogeneity tends to be reduced in models with continuous movements of individuals within patches (Rohani *et al.* 1994*a*), and that in some particular continuous-time models the effects of HDD and HDI can be completely different (Murdoch and Stewart-Oaten 1989).

4.6 Other sources of heterogeneity

A particular form of heterogeneity of risk arises when, at any one time, parasitism is confined to part of the host population, with the remaining hosts protected from parasitism in some kind of refuge. These refuges may take several forms; for example: (1) there may be spatial refuges when parasitism is confined to part of the habitat; (2) there may be temporal refuges depending on the timing and overlap of the different host and parasitoid stages; or (3) some host phenotypes may act as a refuge by completely escaping from parasitism. Refuges of these different types, in which some hosts are invulnerable to parasitoid attack at particular times, are likely to be widespread under natural conditions, and there is certainly no shortage of anecdotal, natural history examples falling into these three categories.

While there have been a few laboratory studies aiming to show the impact of refuges in predator–prey and host–parasitoid systems (e.g. White and Huffaker 1969; Sih 1981; Begon *et al.* 1995), detailed evaluations of the role of physical refuges in the field are rare. A notable exception is the study on the California red scale (*Aonidiella aurantii*) and its parasitoids, particularly *Aphytis melinus* (Reeve and Murdoch 1986; Murdoch *et al.* 1989, 1995, 1996b). Murdoch and his colleagues found that scale populations were largely concentrated in the interior of citrus trees (as much as 90% of the adult female scales were found on the bark of the trunk and on the structural branches in the interior of the tree), while parasitism by *Aphytis* was very much higher on the exterior twigs. Removal of the scales from the refuge had no effect on stability, although it did decrease the density of the exposed populations (see Chapter 5, p. 102). These empirical results provide no clear and consistent view on whether or not refuges tend to stabilise interactions. But model results are much clearer: refuges can have a strong stabilising effect on populations, but only under certain conditions. This section reviews some of the dynamical effects arising from different kinds of refuges relevant to host–parasitoid systems.

4.6.1 Spatial refuges

Spatial refuges have conventionally been classified under two headings: (1) constant proportion refuges and (2) constant number refuges. But, as Lynch *et al.* (1998) emphasise, these categories are a convenience for developing simple models, and spatial refuges can take a variety of intermediate forms in which the proportion or number of hosts protected varies with time or under different conditions (their 'dynamic refuge'). Here we start with the simple case of a constant proportion of hosts within a fixed refuge of some kind; in other words, in each generation there is a fraction of hosts exposed to parasitism (ε) and a fraction ($1 - \varepsilon$) within a refuge. This is therefore a limiting case of the general aggregation model (4.4) in which there are just two host patches with all parasitoids aggregating in one of them. To this we add the possibility that refuge and non-refuge areas may also differ in other respects; for example, the host rate of increase may be different in the two areas due to differences in resource quality (e.g. Prestidge and McNeill 1983; Kidd *et al.* 1990), affecting, for instance, host body size and hence fecundity (e.g. Hanski 1987). This leads to the following generalised proportional refuge model (Holt and Hassell 1993):

$$N_{t+1} = \varepsilon \lambda_1 N_t f(N_t, P_t) + (1 - \varepsilon) \lambda_2 N_t$$
$$P_{t+1} = c N_t \varepsilon [1 - f(N_t, P_t)].$$
(4.11)

Here λ_1 and λ_2 are the host rates of increase outside and within the refuge, respectively, c is the average number of female parasitoids emerging from each parasitised host and $f(\cdot)$ is the fraction of exposed hosts surviving parasitism, given here by the usual Nicholson–Bailey term, $\exp(-aP_t)$.

The stability properties of proportional refuges have been explored on several occasions (e.g. Bailey *et al.* 1962; Hassell and May 1973; Maynard Smith 1974; Murdoch and Oaten 1975; Chesson 1978; McMurtrie 1978; Holt and Hassell 1993; Krivan 1998). With uniform host reproductive rates ($\lambda = \lambda_1 = \lambda_2$) there is a relatively narrow range of intermediate refuge sizes leading to locally stable populations. Outside this region, too many hosts in the refuge lead to an unchecked host population, while too few hosts in the refuge lead to cycles and chaos (Fig. 4.11(a)).[8]

In contrast to this *constant proportion* refuge, several people have explored situations where a *constant number* refuge would be more appropriate; for

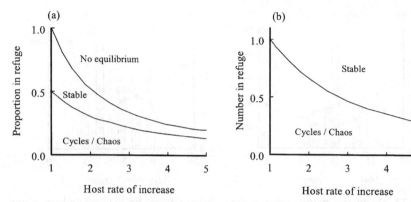

Fig. 4.11 The effect of refuges on the stability of the model (4.11) in relation to the host rate of increase (λ). (a) A constant proportion (ε) of hosts in a refuge. The model is locally stable within the labelled region. Above this region the proportion of hosts protected in refuges is too great for the parasitoids to maintain any equilibrium, and below there are too few hosts protected and the populations show cycles or chaos. (b) A constant number of hosts in a refuge. Stability is now enhanced, although too few hosts in a refuge still leads to unstable cycles. (After Hassell 1978.)

instance, where the refuge habitat is relatively constant and can only support a fixed number of hosts irrespective of their population size (e.g. Maynard Smith 1974; Hassell 1978; McNair 1986; Hochberg and Holt 1995). This situation only differs from that above in $\varepsilon_1 = (N_t - N_0)/N_t$ and $\varepsilon_2 = N_0/N_t$, where N_0 is the number of hosts protected. Constant number refuges are more stabilising than the corresponding constant proportion ones (Fig. 4.11(b)); not surprisingly, since there is now the density-dependent effect of an increasing fraction of the hosts exposed to parasitism as host density increases.

The 'dynamic refuge' model of Lynch *et al.* (1998) differs from these absolute refuges in that the hosts during their development pass into and out of the physical refuge (e.g. the area below the reach of a parasitoid's ovipositor). The refuge effect is thus lessened by high rates of host movement and vice versa. As Lynch *et al.* point out, the process is very similar to one reported by Weisser and Hassell (1996) in which hosts disperse between patches and at any one time may be in transit and therefore temporarily not susceptible to parasitism .

4.6.2 Temporal refuges

Any mismatch in the timing of the searching parasitoid and the incidence of susceptible hosts can be another important source of heterogeneity of risk. This will be most marked in insects with more-or-less discrete generations where vulnerability to parasitism is likely to be over relatively narrow windows of time. The parasitoids' life cycle must therefore closely match that of the hosts if attacker and attacked are to coincide temporally. But even if they are

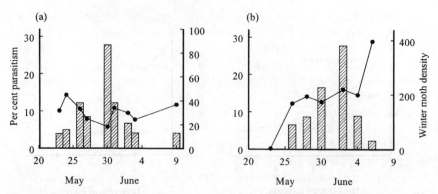

Fig. 4.12 Temporal pattern of parasitism (solid circles) by the tachinid parasitoid, *Cyzenis albicans*, in successive samples of winter moth (*Operophtera brumata*) pre-pupae (hatched bars) descending to the ground to pupate. *Cyzenis* normally emerge well ahead of their winter moth hosts leaving sufficient time to mature a full complement of eggs. In these cases there are no trends in the levels of parasitism during the period of host availability, as shown in (a) for the data in 1966. But in some years (b) parasitoid emergence is relatively late, and the first hosts to appear tend to escape parasitism, as occurred in 1959. (After Hassell 1969*b*.)

present at the same time, host individuals can still experience different risks of parasitism if there is a build-up and decline in parasitoid numbers during the period of host susceptibility. Indeed, this effect will only be totally absent if all hosts, irrespective of their phenology, experience the same density of foraging parasitoids. This would require, first, that all the female parasitoids had emerged and were ready to attack hosts before any of the susceptible hosts appear, and, second, that they survive throughout the period of host susceptibility.[9] In most cases, parasitoid and host emergence will overlap to some degree; two examples are shown in Fig. 4.12, one where parasitism rises through the host period and the other where it is more or less constant.

The effects of phenological asynchrony have been examined in several kinds of host–parasitoid system (e.g. Griffiths 1969*b*; Hassell 1969*a*; Cheng 1970; Munster-Swendsen and Nachman 1978; Munster-Swendsen 1980; Godfray *et al.* 1994). In discrete generation models without age-structure, perfect synchrony has been the automatic assumption, while in fully continuous interactions the question does not arise since all stages are present simultaneously. The first indication that temporal mismatches may act to stabilise host–parasitoid interactions was made by Varley (1947). Subsequently, a number of studies have reported asynchrony from host–parasitoid systems in the field and attempted to include this in population models. For example, Griffiths (1969*b*) and Griffiths and Holling (1969), studying the parasitoids of the European pine sawfly (*Neodiprion sertifer*), described the observed asynchrony by modifying the negative binomial host–parasitoid model to make parasitoid searching efficiency decline as asynchrony increased. Secondly, Munster-Swendsen and Nachman (1978) included temporal asynchrony explicitly in their simulation model of the interaction between the spruce tortricid, *Epinotia tedella* and its ichneumonid parasitoid, *Lissonota* (*Pimplopterus*) *dubius*. In both these cases, asynchrony is thought to have had some stabilising effect on the interacting populations.

A more general treatment of the dynamical effects of asynchrony is given by Godfray *et al.* (1994). Let us consider a univoltine host–parasitoid interaction in which all hosts at the beginning of the season are in a non-susceptible stage (e.g. eggs) and all parasitoids in a non-searching stage (e.g. pupae). For narrative convenience the hosts are assumed to be susceptible to parasitism throughout the larval stages and the female parasitoids are assumed to search for hosts throughout their adulthood (Fig. 4.13). The model of these life cycles can thus be divided into two parts: (1) a within-generation component that describes the dynamics of emergence, maturation and parasitism; and (2) a between-generation component that relates the numbers of surviving or parasitised hosts at the end of one season to the numbers of hosts and parasitoids, respectively, at the beginning of the next season.[10] This between-generation component is very straightforward: each surviving host has a net reproductive rate of λ and each parasitised host produces one adult female parasitoid in the next season.

The within-generation component is more complex, and makes use of the lumped age–class techniques to be described in Chapter 5 (Nisbet and Gurney 1983; Murdoch *et al.* 1987). The host life cycle is divided into eggs, larvae and

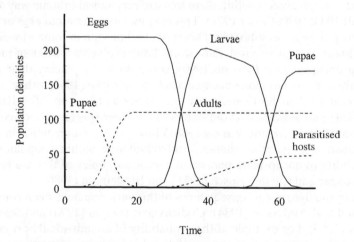

Fig. 4.13 An example of the changes in host (solid lines) and parasitoid (broken lines) life-history stages within one generation. In this case there is no phenological asynchrony since the adult parasitoids complete emergence prior to susceptible hosts (larvae) appearing, and show no mortality while hosts are available. See text for further details. (From Godfray *et al.* 1994.)

pupae, and the parasitoid life cycle into juveniles and adults, as in Fig. 4.13. The full properties of the model are discussed in Godfray *et al.* (1994). When the number of searching parasitoids is constant throughout the period that susceptible hosts are present (and assuming zero handling time), there is no asynchrony and any stability of the interaction must come from other density-dependent processes. As parasitoid emergence is delayed, however, the stability properties change. First, the model becomes increasingly stable as a refuge effect is introduced, up to a point when no other density dependence is needed. Beyond this, if parasitoids do not begin to emerge until late into the period of host susceptibility, then too large a proportion of hosts are protected for the parasitoids to be able to exert any stabilising influence.

In short, delays in parasitoid emergence are another means by which hetero-geneity in the risk of parasitism can be generated; those hosts that emerge early in the host cohort experience a reduction in, or even the complete absence of, parasitoid attack. In certain cases this temporal heterogeneity can promote stability sufficiently to stabilise a model that otherwise would have completely unstable Nicholson–Bailey dynamics.

4.6.3 Phenotypic variability

Variation in the risk of parasitism between host individuals may arise without any heterogeneity in space or time, simply because some individuals are in-herently more or less susceptible to parasitism. For instance, insects generally have a powerful haemocytic defence mechanism that enables them to en-capsulate foreign objects within them that are recognised in some way as 'non-self' (Salt 1963; 1970; Fisher 1971). This may include parasitoid eggs or larvae, especially if they are in suboptimal hosts or are damaged in some way. Because of this threat to their survival, parasitoids have evolved a variety of means of avoiding encapsulation (Vinson 1990); some do so by placing their eggs in tissues away from marauding haemocytes (e.g. Salt 1963, 1968), others are able to incapacitate their host's defences (e.g. Stoltz and Vinson 1979; Rizki and Rizki 1984) and yet others avoid detection by their special surface coat (e.g. Rotheram 1967; Lewis and Vinson 1968; Vinson 1990). Aggregation in the risk of parasitism will thus arise whenever individual hosts within a population vary in this ability to encapsulate parasitoids; some examples of this are reviewed by Messenger and van den Bosch (1971) and Bouletreau (1986).

Some of the dynamical consequences of this variation have been considered by Hassell and Anderson (1984), Godfray and Hassell (1990) and Sasaki and Godfray (1999). For example, if the probability of an individual host escaping parasitism by encapsulation is assumed to be constant, irrespective of how many parasitoid larvae are within the host, the host–parasitoid model reduces exactly to the simple proportionate refuge model above where the size of the refuge is determined by the proportion of hosts that are resistant to parasitism. Such 'all-or-none' encapsulation can thus help to stabilise an otherwise unstable interaction by virtue of this refuge effect. Additional heterogeneity in the form of random variability between individuals in their ability to encapsulate parasitoids does not enhance this effect at all; the size of the refuge is now merely the mean of the probability density function used (Godfray and Hassell 1990).

A rather different picture emerges where the probability of a host surviving by encapsulating the parasitoids is dosage-dependent (i.e. the probability decreases with parasitoid load). Explicit account must now be taken of hosts that are attacked once, twice, thrice and so on. Assuming, for convenience, a Poisson distribution of parasitoid attacks and a fixed probability, η, of any parasitoid egg or larva being encapsulated, survival from parasitism, f, in the general model (4.3), is given by $f(P_t) = \exp[-aP_t(1 - \eta)]$. This is the Nicholson–Bailey model, except that parasitoid searching efficiency is now reduced to $a(1 - \eta)$. Such dosage-dependent encapsulation therefore has no effect on the stability properties of the interaction. This changes, however, if individual hosts vary in their ability to encapsulate. Now, an otherwise unstable Nicholson–Bailey model can be completely stabilised via this mechanism provided λ is relatively small.

Variation in the ability of hosts to encapsulate parasitoids is only one of many ways that host phenotype can affect an individual's risk of parasitism (see Godfray 1994 for a review). Some hosts, for example, strive to avoid parasitism by threat display or crypsis, which may be more or less effective in different individuals. Others mount a physical defence against approaching female parasitoids, by wriggling, dropping off leaves on silken threads or falling to the ground, and these responses are also likely to vary in vigour between individuals. With such heterogeneity arising from so many sources, phenotypic variability may well be a significant component of the overall heterogeneity that permeates host–parasitoid interactions.

4.7 Summary

Heterogeneity in the distribution of parasitism is defined specifically in terms of the distribution of the risk of parasitism amongst the individual hosts in a population. If the distribution is even there is no heterogeneity, as in the Nicholson–Bailey model. As the distribution becomes more aggregated so heterogeneity increases. For example, one way of obtaining the negative binomial model of parasitism is when the distribution of risk is gamma-distributed amongst the hosts.

One obvious way that heterogeneity can arise is by the non-random distribution of parasitism amongst hosts distributed in discrete patches. From the available empirical literature these distributions may be directly or inversely correlated with host density per patch or may be density-independent. Such spatial heterogeneity can therefore be categorised as *host density-dependent heterogeneity* (HDD) (positive or negative), or *host density-independent heterogeneity* (HDI).

A number of models have been explored with spatially explicit distributions of parasitism. Some of these—'fixed aggregation models'—have a single episode of spatial redistribution of the parasitoids at the beginning of the season, after which the patterns do not change until the following generation. Others—'continuous redistribution models'—allow the parasitoids to move from patch to patch within the season, responding to the changing pattern of host exploitation. Any optimal foraging model falls into the latter category. In the case of fixed aggregation, both HDI and HDD parasitism contribute to heterogeneity of risk and are therefore stabilising processes within discrete generation models. If there is continuous redistribution of parasitoids, however, the picture is somewhat different. The role of HDI parasitism is unaltered since the heterogeneity of risk is unaffected by parasitoid movement between patches. HDD parasitism, however, leads progressively to more even patterns of parasitism amongst the patches. Since stability depends on there being sufficient heterogeneity of risk at the *end* of the searching period, the stabilising effect of HDD parasitism will decline the more parasitoids there are and the longer that they search.

The stabilising effect of HDI and HDD (both direct and inverse) can be evaluated using the approximate '$CV^2 > 1$ rule'. This states that a discrete-generation, host–parasitoid interaction should be stable if the square of the coefficient of variation of the distribution of searching parasitoids per patch is roughly greater than one. When applied to field data on the spatial patterns of parasitism from patch to patch, the estimated value of CV^2 was found to be greater than one in about 25% of the cases, and in almost all of these it was the HDI parasitism that had much the greater effect.

Heterogeneity can also arise in other ways. For example, an extreme form of spatial heterogeneity arises when some hosts are completely protected from parasitism within a physical refuge. There may now tend to be either a constant proportion or a constant number of hosts escaping parasitism in each generation. Alternatively, there may be temporal refuges in which some hosts escape parasitism due to the relative timing of susceptible hosts and searching parasitoids, or there may be variability in host behaviour or physiology which affect an individual's chances of being parasitised. In all these cases, enhanced heterogeneity of risk is a stabilising influence on the interacting populations.

Notes

1. Note that the term $\ln(N/S)$ is the k-value for mortality described by Varley and Gradwell (1960), except for the use of natural instead of common logs.
2. For some parameter combinations the interaction is also unstable for very low values for λ (Hassell and May 1973; Rohani *et al.* 1994*b*).
3. Parasitoids and time are interchangeable here; thus keeping parasitoid density

constant, but varying the time of search would produce the same apparent inter-ference relationships.

4. This assumes, of course, that the 'attractiveness' of the different patches to para-sitoids remains constant throughout the season; otherwise, if the redistributing parasitoids entered patches at random, the heterogeneity of risk would decline as the season progresses.

5. The criterion is expressed in terms of CV^2 rather than CV because of the way it appears naturally in eqn (4.10).

6. More specifically, if there is a total of y hosts in z patches, and the distribution of searching parasitoids per patch is given by q_j ($j = 1, 2, ..., z$), and we let p_i be the density of searching parasitoids in the vicinity of the ith individual host ($i = 1, 2, ..., y$), it is the CV^2 of the p_i-values, rather than the q_j-values, that is used.

7. Cronin and Strong (1990) also found $CV^2 > 1$ from the distribution of parasitism in the case of another *Prokelisia* parasitoid, the mymarid *Anagrus sophiae*.

8. This picture is somewhat complicated if the hosts inside and outside the refuge have different reproductive rates (Holt and Hassell 1993). First we note that in the absence of any parasitoid aggregation these unequal values of λ have no effect in promoting stability. But once parasitoids aggregate in some patches over others, we need to consider the combined terms $\varepsilon\lambda_1$ and $(1 - \varepsilon)\lambda_2$. In particular, spatial hetero-geneity in host rates of increase can enhance the stabilising effect of heterogeneity in parasitism rates, provided that parasitism is heavier on those hosts with the higher rates of increase.

9. An unlikely third possibility is that parasitoid emergence and mortality were balanced such that a more-or-less constant density of parasitoids were searching throughout this period.

10. The structure of this model is thus similar to that of Rohani *et al.* (1994a) involving discrete generations but continuous within-generation movement.

5

Continuous time and age-structure

5.1 Introduction

Most of the host–parasitoid systems explored so far in this book have involved coupled interactions with discrete and synchronised generations. Such discrete-generation life cycles are most frequently found in univoltine species in temperate regions where seasonality often synchronises populations and provides a natural interval between the appearance of successive generations. Seasonality also often synchronises the appearance of the first generation of multivoltine species within a year, although successive generations throughout the season tend to show progressive overlap of their life-cycle stages. Focusing on discrete-generation life cycles does not mean that age-structure is necessarily neglected in these models. For example, the combination of parasitoids and additional host density dependence requires specific assumptions about the order with which these occur in the host's life cycle, which is then reflected in the model structure (Chapter 3, p. 46). Another example, in this case with an explicit treatment of age-structure, was discussed in Chapter 4, p. 90, involving the stabilising effects of phenological asynchrony between hosts and parasitoids within a single generation period.

But there are other kinds of interaction in which host and parasitoid life cycles overlap so much that a quite different model structure is required. For example, some insects, particularly in warmer climes, breed more-or-less continuously for much of the year; this results in considerable overlap of generations with all stages tending to be present at the same time. For such systems, continuous time models are more appropriate, often with an explicit treatment of the age-structure of the interacting populations. Several of these models have been very useful in illustrating how the details of host and parasitoid life cycles can have important dynamical effects. In this chapter the host–parasitoid interactions are framed in continuous time, and the main emphasis is on the interesting dynamics that appear as a direct result of age-structure. Age-structure tends to introduce time delays, and the effects of these are first illustrated within the familiar Lotka–Volterra model for predator–prey systems. We then turn to interactions with more detailed age-structured effects that are explored using delay-differential equations.

5.2 The Lotka–Volterra model

The dynamical effects of natural enemies in continuous time are usually explored using variants of the basic Lotka–Volterra equations (Lotka 1925; Volterra 1926). The assumptions are straightforward. In the absence of predators, the identical prey, N, increase exponentially at a rate r. All predators, P, are also identical and eat prey at a rate aPN, where a is the attack rate; the implicit functional response is therefore linear, as in the Nicholson–Bailey model. In the absence of prey, the predators dwindle exponentially through density-independent mortality at a rate d, but any prey eaten are converted into new predators at a rate $caPN$, where c reflects the predators' efficiency at converting prey eaten into offspring. The model can therefore be written as:

$$\frac{dN}{dt} = rN = aPN$$
$$\frac{dP}{dt} = -dP + caPN. \tag{5.1}$$

The properties of this model are very well known—the populations show neutrally stable cycles whose amplitude depends only on the initial population levels. This neutral stability is a special case representing a 'knife edge' between regions of expanding and dampened oscillations. Anything, therefore, that adds stability or instability moves the populations into one or other of these two regions, and many variants of the Lotka–Volterra model have demonstrated this (see Gotelli 1995 for an excellent review). For example, if the prey population in the absence of predation shows a logistic growth rate, the prey equation in model (5.1) becomes:

$$\frac{dN}{dt} = rN\left(1 - \frac{N}{K}\right) - aPN \tag{5.2}$$

where K is the carrying capacity, and the populations now always move to a stable equilibrium. Stability will also occur if the linear functional response above is replaced by a type III response, or if density dependence is introduced via the predation term—for instance, if the predators interfere with each other, in which case we may have:

$$\frac{dP}{dt} = -dP + aNP^m \tag{5.3}$$

where $m > 0$ is the measure of density dependence. Once again, the populations will move to a stable equilibrium.

Like the Nicholson–Bailey model described in Chapter 2, the basic Lotka–Volterra model is a very simple representation of a basic resource–consumer system that illustrates the intrinsic propensity for such systems to oscillate. A disadvantage it has, however, is the complete lack of any time lags, developmental or otherwise, that are inevitable in all such interactions. But this can easily be remedied as illustrated in the next section.

5.3 The destabilising effect of time lags

A number of authors have explicitly included age-structure in the Lotka–Volterra model by restricting predator attack to only one prey stage, juveniles or adults (e.g. May 1974b; Smith and Mead 1974; Hastings 1983, 1984). Here we consider a straightforward extension of the Lotka–Volterra model described by Godfray and Chan (1990), in which the time taken to grow from oviposition to the adult stage for both hosts and parasitoids (τ_N and τ_P, respectively) is incorporated explicitly:

$$\frac{dN_t}{dt} = rN_{t + \tau_N} - N_t f(P_t) - \mu_N N_t$$

$$\frac{dP_t}{dt} = N_{t + \tau_P} f(P_{t + \tau_P}) - \mu_P P_t. \tag{5.4}$$

In the host equation, the first term is the recruitment to the adult host population, which depends on births τ_N time units ago (r now takes both fecundity and juvenile mortality into account). The next two terms are mortalities that the adult hosts suffer: first, parasitism at a rate $f(P_t)$ and second, density-independent mortality from other sources at a rate μ_N. In the parasitoid equation, the first term is the recruitment to the adult parasitoid stage, which depends on the number of hosts parasitised τ_P time units ago, and the second term is the mortality of the adult parasitoids at a constant rate μ_P. Godfray and Chan introduced density dependence in the parasitoids' attack rate by assuming $f = aP^m$ as in eqn (5.3). Alternatively, parasitoid density dependence has sometimes been introduced via:

$$f = k \ln\left(1 + \frac{aP}{k}\right) \tag{5.5}$$

where the parasitoid attack rate decreases with the density of searching adult parasitoids depending on the value of k (density dependence gets stronger with lower k and is absent when $k \rightarrow \infty$) (Godfray and Hassell 1989; Godfray and Waage 1991; Gordon *et al.* 1991; Reeve *et al.* 1994a). The stability properties of this model are shown in Fig. 5.1 in terms of the value of k (normalised) needed for the interaction to be stable as the time lags, τ, increase. To the left of the graph, where there are no time lags ($\tau = 0$) and $k \rightarrow \infty$, the neutrally stable Lotka–Volterra model is recovered. As τ increases the line shows the critical value of k falling as the amount of parasitoid density dependence needed for stability increases. Thus, increasing time lags reduce the stable area until, with a time lag of exactly one generation interval, the

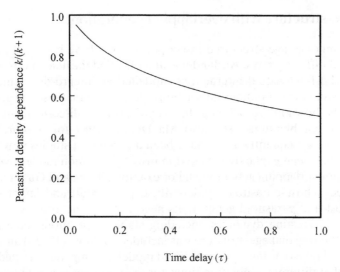

Fig. 5.1 Plot of the stability conditions from model (5.4) in terms of the normalised amount of parasitoid density dependence (k) and the time delays (τ) for host and parasitoid development assuming that these are equal, $\tau_N = \tau_P$. The line indicates the minimum amount of density dependence needed for the interaction to be stable, as explained further in the text. The Lotka–Volterra model is recovered to the left of the graph, when $\tau = 0$, while the discrete-generation negative binomial model is recovered to the right of the graph, when $\tau = 1$. If the assumption of $\tau_N = \tau_P$ is relaxed, particularly if the parasitoid generation time is half that of the host, the interactions become cyclic with a period of roughly one host generation (Godfray and Chan 1990). This is further discussed in Section 5.4, p. 102.

discrete-generation negative binomial model of May (1978) is obtained in which stability requires $k < 1$. Using eqn (5.5) to describe how the instantaneous risk of parasitism increases with parasitoid density is justified as follows. For $k \to \infty$ the risk increases linearly, but as k decreases in magnitude the risk tends more and more to asymptote as parasitoid density increases. Numerous functions could have been used to describe this relationship, but this particular function is chosen because on summing the instantaneous risk over the period of host susceptibility (assumed to be one time unit) the probability of avoiding parasitism is $(1 + aP/k)^{-k}$, exactly the function used in the negative binomial model of May (1978). As seen in Chapter 2, this discrete-time model is stable for $k < 1$, and the equivalent continuous-time model (with a time lag) is also stable for $k < 1$. This simple model therefore emphasises that it is the inclusion of a time lag that is destabilising, and not the transition from continuous to discrete time *per se*. It also explains why the Nicholson–Bailey model is more unstable than the equivalent Lotka–Volterra formulation (May 1973).

5.4 Age-structure with overlapping generations

The dynamics of age-structured insect populations have been examined in a number of different ways. Auslander *et al.* (1974) led the field in this with their study of host–parasitoid interactions modelled as integro-differential equations. Despite the continuous interactions, they found that discrete generation 'waves', as shown in Fig. 5.3 on p. 107, could easily be discerned with suitable parameter combinations (see also MacDonald 1989 for a more general treatment). Subsequently, there have been a number of studies in which age-structure has been explicitly included in insect population models with either discrete or overlapping generations. For example, Wang and Gutierrez (1980) developed a host–parasitoid model with explicit adult and larval stages in which host and parasitoid generations need not be synchronous or of equal duration. Interesting dynamics, including stable populations, emerged as a result of the way that age-structure was included. Bellows (1982b) and Bellows and Hassell (1984), using a parameterised model of competing bruchid beetles, showed that different generation times for the two species was the overriding factor in determining the outcome of competition. Hochberg and Waage (1991) developed an age-structured, continuous-time model for populations of the rhinoceros beetle (*Oryctes rhinoceros*) that are limited by density-dependent larval mortality and a baculovirus. Their model simulations agree with long-term field observations on the depression of the beetle numbers brought about by the pathogen. And Hassell *et al.* (1999) developed an age-structured model for the interaction of spruce budworm with the development of the forests in north-eastern Canada, showing the interplay of the insect dynamics with a much longer cycle of foliage development.

The most obvious way to deal with age-structure in population dynamics is to treat age as a continuous variable and to phrase models as systems of partial differential equations with boundary conditions. This gives rise to the classic Kermack–McKendrick (1927) model and its variants. While mathematically elegant, the problem with this approach is that partial differential equations are hard to deal with, both analytically and numerically. Fortunately, the insect life cycle offers a simpler alternative. Most of the major age-differences in demographic parameters occur between major life stages (egg, larvae, pupae, adults) or between larval instars. Often, to a good approximation, variation within these stages can be ignored. The partial differential equations can then be rephrased in more tractable form as integro-differential equations, systems of ordinary differential equations or, the most frequently used trick in the parasitoid literature, as delay-differential equations.

This whole approach has been extensively developed by R. M. Nisbet, W. S. C. Gurney and colleagues, first in the context of single species (Nisbet and Gurney 1982; Blythe *et al.* 1983; Gurney *et al.* 1983; Blythe 1984) and subsequently for host–parasitoid (e.g. Gordon 1987; Murdoch *et al.* 1987; Godfray and Hassell 1989; Briggs *et al.* 1993) and host–pathogen systems (Briggs and

Godfray 1996). Their delay-differential equations for insect life cycles assume that individuals can be 'lumped' into distinct classes, within which all experience the same demographic processes that vary from stage to stage. A variety of developmental schedules can be assumed (e.g. weighted maturation times) for development and survival, and these do not necessarily need measures of uncertainty (Blythe *et al.* 1983). In their simplest, single-species model only two stages, reproductive adults and juveniles, were considered. The adult dynamics depend on their death rate and the recruitment rate from the juveniles; the juvenile dynamics depend on the birth rate and the maturation rate into new adults. The interesting property of these models is the ease with which persistent fluctuations arise (stable cycles or chaos) with a period and amplitude not dissimilar to the generation cycles often observed from laboratory-maintained insect populations, such as blowflies (Nicholson 1950, 1954, 1957; Gurney *et al.* 1983), bruchid beetles (e.g. Utida 1941; Hassell *et al.* 1989; Shimada and Tuda 1996) and stored product Lepidoptera (Gurney *et al.* 1983; Gordon *et al.* 1988; Hassell *et al.* 1989; Begon *et al.* 1995). A thorough analysis of mechanisms causing these cycles has been given by Gurney and Nisbet (1985), and also by Briggs *et al.* (2000). Their models indicate generation cycles arising from asymmetric competition between different-aged larvae in which the young larvae are more susceptible to shortage of food but exert weaker competitive effects. Cannibalism of eggs by older larvae also seems essential to generate cycles with the period observed in laboratory cultures.

5.4.1. The California red scale

In an important paper, Murdoch *et al.* (1987) extended this approach by developing a model framework for stage-structured, host–parasitoid interactions in continuous time. The work was largely motivated by a long-term study of the red scale (*Aonidiella aurantii*) and its aphelinid parasitoid, *Aphytis melinus*, in Southern California. Following its introduction into California between 1868 and 1875, the red scale became a major pest of citrus trees (DeBach *et al.* 1971). *Aphytis melinus*, which is currently the most effective parasitoid, was introduced into California between 1956 and 1957, and appears to have been responsible for the dramatic reduction of scale population densities which have subsequently been maintained at these low levels, making the red scale an example of highly successful biological control. A striking feature of the red scale is the apparent stability of the populations around a low equilibrium level, explanations for which have been sought in a series of papers (e.g. Reeve and Murdoch 1985, 1986; Murdoch *et al.* 1987, 1989, 1996b).

The life cycle and natural history of the red scale are well known (e.g. Reeve and Murdoch 1985, 1986; Murdoch *et al.* 1989, 1996). Although reproduction and development of the red scale are continuous and the life cycle can be completed in about 5 weeks (at 28 °C), there are only about 3.5 generations per year due to their relative inactivity over the winter months. The *Aphytis melinus* primarily attack the second and third instars and develop much more rapidly than the red scale (about three generations to each one of the red scale). Red scale populations are found throughout the citrus trees, but the density of scales tends to be much higher on wood in the interior of the trees, while parasitism is much higher in the outer parts of the canopy. The woody interior branches therefore appear to be a refuge from parasitism by *Aphytis*. In an interesting field experiment lasting three generations and involving removal of the refuge population, Murdoch *et al.* (1996) found, if anything, a decrease in temporal variability. Although this was a short period in which to detect effects on stability, the results suggest that the refuges are not of paramount importance to the persistence of the red scale populations.

Murdoch *et al.*'s (1987) basic model addresses, in particular, the dynamical effects of having a relatively long invulnerable host stage that is free from parasitoid attack (see Briggs *et al.* 1999a for a review). Following the techniques of Gurney *et al.* (1983), the models are composed of a set of balance equations for adult and unparasitised juvenile hosts, juvenile parasitoids and searching adult parasitoids. For each stage there is a fixed developmental period and an instantaneous density-independent mortality. Thus hosts leave the susceptible juvenile stage by maturing to adults, by being parasitised or by succumbing to the density-independent mortality. Those entering the adult stage then lay eggs at a constant rate per day and also die at a constant rate. Parasitism occurs by the adult parasitoids encountering susceptible hosts randomly at a constant rate (i.e. with a linear functional response). These juvenile parasitoids then develop into searching adults after a fixed developmental delay.

Murdoch *et al.* (1987) discuss a number of limiting cases in this model which are important for understanding its properties and relationships to other models. First they assume that adult hosts have a constant reproductive rate but an infinitesimally short invulnerable adult stage. With this assumption of instantaneous host reproduction, the model always has a propensity for oscillations with a period of one-host generation, or harmonics of that interval. However, their numerical studies with adult hosts living for a finite time, showed a tendency for dampening of these cycles which were therefore not considered further. Another limiting case arises if, in addition, the parasitoid development time is instantaneous. Since the red scale population grows exponentially in the absence of parasitism and the parasitoids respond instantaneously with random parasitism and a linear functional response, it is not surprising that Lotka–Volterra-like, neutrally stable cycles occur in this case. Once any parasitoid developmental period is introduced, instability is added and the neutral cycles give way to an unstable interaction.

Introducing an invulnerable host stage into these models always contributes to stability. This effect is modest if it is just the juvenile hosts that are immune —only for a narrow range of invulnerable juvenile periods can this counteract the destabilising tendency of the juvenile parasitoid's developmental period.

The effect, however, can be much more marked if it is the adult hosts that are immune. In particular, stability is enhanced by high adult longevity compared to that of the juvenile hosts, and by relatively short parasitoid developmental rates. When the model is parameterised for the red scale, however, the dynamics lie outside the stable region, and limit-cycles occur of much greater amplitude than observed in the field. Therefore, Murdoch *et al.* (1987) concluded that while continuous generations and the invulnerable adult stage can probably be a significant cause of stability, these alone are probably insufficient to be the primary cause for the observed stable populations.

A broadly similar effect of having long-lived adults was observed by Hastings (1984) from an age-structured version of the Nicholson–Bailey model (applied to wolves and ungulates!). Here only juveniles are preyed upon and the longevity of the adults is governed by a constant survival rate parameter. He showed that the age of adult senescence was a critical factor in moving the dynamics from Nicholson–Bailey expanding oscillations to a locally stable interaction with some degree of overlap of generations. This works by the long-lived, invulnerable adult stage counteracting the diverging oscillations by providing a refuge for prey to 'ride out' the periods of high predator density. In particular, the presence of even a very small number of long-lived adult prey individuals can be an important stabilising mechanism due to this 'storage effect'.

There are several other interesting features of host–parasitoid interactions illustrated by the red scale system. For example, parasitoids of the red scale are influenced by the size of their hosts in a number of interesting ways. Thus, they host-feed[1] rather than oviposit on the smallest scales, their clutch sizes increase with the size of the parasitised scale and the sex ratio of the parasitoid progeny becomes increasingly female-biased as scale size increases. Female parasitoids are therefore most effective in producing females for the next generation when attacking the larger hosts. The dynamical effects of each of these have been explored using developments of the basic model above (reviewed in Murdoch *et al.* 1997; Briggs *et al.* 1999*a*). Size-selective host-feeding (SSH) was explored by Murdoch *et al.* (1992*b*) by dividing the vulnerable immature stages into a younger group (of duration T_0) which were fed on by parasitoids (at rate a_0) and an older group (of duration T_1) which were parasitised (at rate a_1). The original invulnerable stage model is thus recovered when there is no host-feeding stage (i.e. $T_0 = 0$). The greater the degree of host-feeding relative to parasitism (i.e. as a_0T_0/a_1T_1 increases), the more time delays are introduced in the action of the parasitoids. As the parasitoids host-feed more and parasitise less there is, at first, a decrease in the region of cycles in the basic model and a consequent increase in the region of stable equilibria. But further increases in host-feeding, and therefore more time delays, lead to a different kind of instability via large-amplitude cycles which can greatly reduce the region of stability.

The effects of the parasitoids' clutch size being host size-dependent (SCH) can be explored from a similar model in which the invulnerable hosts are again divided into two; the smaller group receiving only one parasitoid egg on parasitism, while the larger group receives more than one egg (Murdoch *et al.* 1997). Thus the basic model is recovered if the parasitoid attack rate for both groups is the same, if they last for the same duration and if one egg is laid at each encounter with a host in either group. Increasing the number of parasitoid eggs laid per encounter with the larger hosts increases the impact of the parasitoids, and hence their delayed density-dependent effect on the hosts, and so has a similar effect to the parasitoids in the SSH model.

Several further refinements were considered by Murdoch *et al.* (1997). For example, combining the two effects above (SCH plus SSH) within the same model (i.e. the vulnerable hosts are divided into three groups), further increases the delayed impact of the parasitoids and hence the region of 'delayed feedback cycles'. They also consider the parasitoid's own state, as well as the size of encountered host. For example, as observed for *Aphytis* (e.g. Luck *et al.* 1982), parasitoids may preferentially feed, rather than oviposit, on small hosts, but the probability of this happening decreases as the parasitoid's current egg load increases. Now the basic model is recovered as a limiting case if the large and small host stages are fed on with equal probability for a given parasitoid egg load. Once the effect of increasing parasitoid egg load in reducing the probability of host-feeding is greater for large rather than small hosts, the delayed effect of the parasitoids increases, and this again tends to cause the cycles observed in the SSH and SCH models above.

5.4.2. Generation cycles

Scale insects have relatively long-lived adult stages compared to other kinds of insects. The models from the previous section demonstrate the potential importance for stability of having such long host stages that are invulnerable to parasitism. However, most insects (e.g. Lepidoptera, Diptera, Homoptera), have much shorter adult stages relative to the length of the immature stages exposed to parasitism, and therefore the invulnerable host stage contributes less to stability. A related set of models has been developed around these life cycles, in which the stabilising effect of age-structure is judged in relation to the minimal amount of density dependence in the parasitoid attack rate that is needed for stability.

Let us assume that the instantaneous risk of parasitism in such models is described by eqn (5.4). In contrast to the models with random parasitism and long-lived invulnerable adult hosts, stability now arises from both the age-structure and the processes, described in previous chapters, that make the parasitoids' attack rate density-dependent. For example, Godfray and Hassell (1989) considered the idealised host and parasitoid life cycles shown in Fig. 5.2. Following Murdoch *et al.* (1987), each stage is assumed to suffer a density-independent mortality at an instantaneous rate μ_x (where x is the stage in

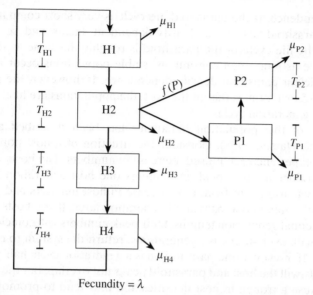

Fecundity = λ

Fig. 5.2 Idealised host and parasitoid life cycles assumed in the model of Godfray and Hassell (1989). The host life-cycle consists of three immature stages, H1, H2 and H3, of which H2 is susceptible to attack by the parasitoid. The adult host stage, H4, produces eggs, H1, at a rate λ. The parasitoid has an immature stage, P1, and the adult searching stage, P2. The probability of a host being parasitised is governed by a function $f(P)$. All stages suffer density-independent mortality indexed by the subscripted variables, μ, and the duration of each stage is given by the subscripted variables, T. (From Godfray and Hassell 1989).

question) such that the probability of surviving until the end of a stage x lasting for T_x units is given by $\exp(-\mu_x T_x)$. Having specified these divisions of the life cycle, the system can be represented by balance equations for the rate of change of hosts and parasitoids through their respective stages (full details are given in Godfray and Hassell 1987, 1989).

There are three distinct regions of stability. If there is relatively weak density dependence in the parasitoid attack rate, there is an unstable region showing typical divergent host–parasitoid cycles. If, however, there is marked density dependence in the parasitoid attack rate, there are two distinctive regions. First, the populations may be stable with constant population sizes and all host and parasitoid age classes present at the same time. Second, the local equilibria may be unstable, showing stable cycles of host and parasitoid populations—with the periods of both being roughly equal to the average duration of one host generation, and therefore giving the appearance of more-or-less discrete generations. Which of these outcomes—'generation cycles' or stable, overlapping generations—occurs depends largely on the ratio of the lengths of the host and parasitoid life cycles, and the degree of parasitoid density dependence. If the parasitoid life cycle is very short compared to the host, the parasitoid acts as a density-dependent factor and so enhances stability. If the life cycle of the parasitoid is roughly the same as that of the host, or twice as long, then continuous, stable populations occur as long as there is sufficient parasitoid density dependence. If, however, the parasitoid life cycle is 0.5 or 1.5 times that of the host, the populations are likely to break into discrete generation cycles.

The basis of this population behaviour has been described as follows (Godfray and Hassell 1988). Consider the situation of a host population at equilibrium experiencing a 'pulsed' increase in numbers. The hosts will reproduce and produce another peak in numbers one host generation later. The parasitoids will also profit from this increase in host numbers and produce a peak in their numbers one parasitoid generation later. If the hosts and parasitoids have equal generation lengths, their peak numbers will coincide and the parasitoids will, over successive generations, return the system to continuous generations. If, however, the parasitoid has a generation cycle half that of the host, not only will the host and parasitoid peaks not overlap, but the parasitoid peak will cause a trough in host densities and thus tend to promote discrete generation cycles.

A partially parameterised version of a similar model, in which the parasitoid developmental lag was influenced by the host stage, is given by Gordon *et al.* (1991) for a laboratory host–parasitoid interaction between the stored product moth, *Ephestia* (= *Cadra cautella*), and its ichneumonid parasitoid, *Venturia canescens*. The estimated ratio of parasitoid to host developmental periods is 0.4, indicating that generation cycles will occur provided the density dependence in the parasitoid attack rate is strong enough.

These results raise the intriguing possibility that parasitoids may be driving the apparent 'generation cycles' that are sometimes seen in tropical host–parasitoid interactions (Fig. 5.3(a)). The models suggest that three requirements need to be met:

(1) the parasitoid generation times must be roughly 0.5 or 1.5 times that of the host;
(2) the invulnerable adult host stage must be relatively short; and
(3) there must be some form of density dependence in the parasitoid's attack rate.

There is a difficulty, however, in drawing this conclusion more strongly, since transient cycles can occur when stable populations with continuous generations are sufficiently perturbed, and these too will tend to be of roughly a one-generation period. The best evidence, therefore, for dynamically driven generation cycles comes from some laboratory host–parasitoid interactions where generation cycles are clearly discernible in uniform environments long after any perturbations arising from the starting conditions of the experiment (Fig. 5.3(b)). Encouragingly, these laboratory examples all come from inter-

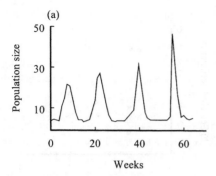
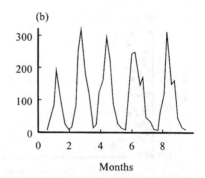

Fig. 5.3 Examples of generation cycles in field and laboratory insect populations. (a) Numbers of eggs per palm of the leaf-mining hispine beetle, *Coelaenomenodera elaeidis*, on oil palm in West Africa between 1970 and 1971 (Mariau and Morin 1972). (b) Numbers of larvae of the flour moth, *Anagasta kuhniella*, in a controlled-environment room in which they are attacked by the ichneumonid parasitoid, *Venturia canescens* (Hassell and Huffaker 1969).

actions where the parasitoids have generation periods of approximately half those of their hosts.

The salt-marsh planthopper

The role of parasitoids in producing generation cycles in the field has been explored by Reeve *et al.* (1994*a,b*) in an interesting study that, like the California red scale, demonstrates how readily these stage-structured models can be applied to specific host–parasitoid interactions with overlapping generations. The salt-marsh planthopper (*Prokelisia marginata*) feeds on cordgrass (*Spartina alterniflora*) along the Florida coast, and is attacked in the egg stage by a mymarid parasitoid (*Anagrus delicatus*). The parasitoid life cycle is substantially shorter than that of the host, taking about 28 days from egg to adult compared to 42 days for the host (at 25 °C). Although host reproduction and parasitism take place throughout the year, the planthopper populations typically tend to show distinct 'generation cycles' within each year.

Of the three requirements for 'generation cycles' suggested above, the planthopper system certainly shows two of them: reproduction is more-or-less continuous and the adult stage is relatively short. The third requirement of a density-dependent parasitoid attack rate was examined in some detail by Reeve *et al.* (1994*a*) and two possible mechanisms found. First, there is clear evidence of highly heterogeneous rates of parasitism across host patches (Stiling and Strong 1982; Strong 1989; Cronin and Strong 1990). This density-independent heterogeneity (HDI) is illustrated in Fig. 5.4(a) and has been quantified using a gamma-distribution of the risks of parasitism across patches (with aggregation parameter k) and estimating k for each of the 29 weeks of

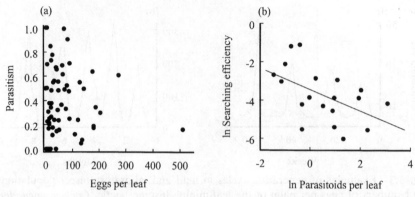

Fig. 5.4 (a) Relationship showing host density-independent proportions of hosts parasitised by the mymarid parasitoid, *Anagrus delicatus*, on the eggs of the salt-marsh planthopper, *Prokelisia marginata*. (b) The decline in *Anagrus* searching efficiency as parasitoid density per leaf increases. (From Reeve *et al.* 1994*a*.)

the study (Reeve *et al.* 1994*b*). The mean value of $k = 0.48$ indicates a highly heterogeneous distribution of risks of parasitism. The likely cause of this, rather than reflecting uneven distributions of searching *Anagrus*, was thought to be external factors, such as tidal vulnerability, which varied significantly between patches. A second mechanism for generating density-dependent parasitoid attack rates was found by placing laboratory-generated host patches in the field and allowing these to be parasitised by the natural *Anagrus* population, whose density was estimated using sticky traps (Cronin and Strong 1990, 1993). The resulting interference, which may either be due to direct interference between adults or to 'pseudointerference' reflecting the hetero-geneity of risk between patches, is illustrated in Fig. 5.4(b).

Reeve *et al.* (1994*a*) explored the effects of both these kinds of density dependence within stage-structured models of the system. Figure 5.5 shows the stability properties from numerical simulations including the HDI hetero-geneity of parasitism in terms of two free variables: the degree of aggregation in the distribution of parasitism, k, and the daily density-independent death rates of host nymphs. The boundaries are similar to those in Godfray and Hassell (1989) above (lines in Fig. 5.5). For values of k above approximately 1.5 the interaction shows typical host–parasitoid cycles. Below this, there are either 'generation cycles', which arise because of the relative host and parasitoid generation times, or a region of stability that is due to the observed length of the invulnerable adult stage. The average value of k found in the field by Reeve *et al.* (1994*b*) was 0.48, which suggests that *Anagrus* has the potential to cause generation cycles except at very high nymphal mortality rates. Their second model differs in including the linear interference relationship shown in Fig. 5.4(b). The resulting stability boundaries are qualitatively the same as before: with too little interference (parasitoid density dependence) there are

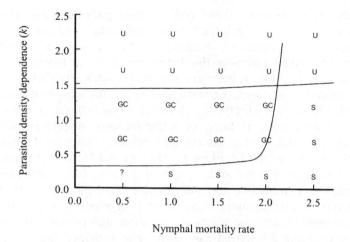

Fig. 5.5 Stability boundaries from a detailed age-structured model of the interaction of the salt-marsh planthopper (*Prokelisia marginata*) and its mymarid parasitoid, *Anagrus delicatus*, in terms of the degree of density dependence in the parasitoid attack rate, *k*, and the amount of density-independent nymphal mortality of the planthopper (Model A of Reeve *et al.* 1994*a*). The letter 'U' denotes a region of large-amplitude, host–parasitoid cycles; 'GC' is a region of generation cycles; and 'S' is a region of local stability. The lines show the corresponding boundaries obtained by applying the model of Godfray and Hassell (1989). (From Reeve *et al.* 1994*a*.)

host–parasitoid cycles. With stronger density dependence, there are either 'generation cycles' or stable interactions, with the latter again being the result of the observed invulnerable adult stage.

A further example where these stage-structured models have been para-meterised for a real system, in this case a multispecies interaction, is given in Chapter 6.

5.5 Summary

Host–parasitoid interactions with discrete, synchronised generations are most frequently found in univoltine species in temperate regions, where seasonality provides a natural interval between the appearance of successive generations. On the other hand, some insects, particularly in warmer regions, breed more-or-less continuously for much of the year leading to considerable overlap of generations. This chapter reviews some continuous-time models for host–parasitoid systems and the ways that age-structure in these interactions can affect dynamics.

The usual framework for continuous-time insect models with age-structure uses the stage-structured approach of R. M. Nisbet, W. S. C. Gurney and colleagues. This has been developed for host–parasitoid interactions by Murdoch *et al.* (1987) and explored both generally and in the context of a long-term study on the population dynamics of the red scale in California. In particular, the model addresses the dynamical effects of having a relatively long-lived host stage that is free from parasitoid attack. Stability is enhanced by these invulnerable stages, particularly when it is the adult host stage that is long-lived. Murdoch *et al.* (1992*b*) also explore the effects of size-selective host-feeding and parasitoid clutch size within the same model framework.

Some insects, such as scale insects, have relatively long-lived adult stages. Other insects, such as the Lepidoptera, Diptera and Homoptera, have much shorter adult stages relative to the length of the immature stages exposed to parasitism. The invulnerable host stage therefore contributes less to stability, making cyclic behaviour more likely. Models in which density dependence in the parasitoid attack rate is introduced, show that population cycles of a roughly one-generation period can easily arise provided that there is: (1) sufficient density dependence; and (2) the parasitoid life cycle is 0.5 or 1.5 times that of the host. If, however, the hosts and parasitoids have more-or-less equal generation lengths, these cycles give way to stable populations with continuous, overlapping generations. Generation peaks are now only visible as transients following disturbance from the equilibrium.

By including realistic life cycles, these models not only reveal the importance of age-structured effects such as invulnerable stages and relative generation times to stability and cycles, they can also be parameterised to correspond well to real systems.

Notes

1. Some parasitoid species both feed from and oviposit in the same host. Others, use different hosts for food and oviposition, in which case host-feeding typically causes the death of the host without parasitoid reproduction.

6

Multispecies host–parasitoid systems

6.1 Introduction

The vast majority of insect hosts, especially in natural habitats, are attacked by more than one parasitoid species, and most parasitoids will attack more than one host species. The population dynamics of natural host–parasitoid systems will therefore not be properly understood without facing up to the complexities of multispecies interactions.

Our current understanding of host–parasitoid population dynamics comes largely from models of two- and three-species interactions. These simple systems correspond quite well to some natural examples in the field in which hosts are effectively attacked by only one or two parasitoid species (e.g. Askew and Shaw 1986; Hawkins and Lawton 1987; Hawkins 1988), and to some of the classical biological-control successes which effectively involve only the pest and the conquering natural enemy. At the other extreme are highly intricate webs of interacting species. For instance, Memmott et al. (1993, 1994) describe a community of Costa Rican leaf-miners and their parasitoids that involves 134 host species attacked by 86 parasitoid species. Webs of such complexity are common amongst host–parasitoid communities (e.g. Askew 1961; Hawkins and Goeden 1984; Askew and Shaw 1986; Rott et al. 1998; Müller et al. 1999). Because they contain so many host species variously 'linked' by specialist, oligophagous[1] and generalist natural enemies, a complete description of the population dynamics of each species in the system is not a feasible way of trying to understand the processes structuring and maintaining the community as a whole. Yet such an understanding of community structure is one of the major challenges in ecology. One approach has been macroecological, in which communities are treated as entire entities, looking for patterns in the abundance and distribution of their constituent species (e.g. Gaston and Lawton 1988; Holt et al. 1997; Lawton et al. 1998; Ruggiero et al. 1998). Alternatively, some aspects of community structure can be predicted from models using very simple and general descriptions of species interactions (Laska and Wootton 1998); for example, community matrices with elements representing the linearised interaction coefficients at the community equilibrium (May 1974b), or systems of Lotka–Volterra equations (e.g. Pimm and Lawton 1977, 1978; Lawton and Pimm 1978; Marrow et al. 1992). Yet another approach has been to develop mechanistic models based on observations of real communities. This has been particularly fruitful in studying plant communities (e.g. Tilman 1982; Pacala and Tilman 1994; Rees *et al.* 1996) and freshwater communities (e.g. Grover 1990; Grover 1991; Holt *et al.* 1994), but it has also had some success in terrestrial animal communities; for example, in understanding guilds of *Drosophila* competing in ephemeral resource patches (Atkinson and Shorrocks 1984; Shorrocks and Bingley 1994).

Recent work on quantifying host–parasitoid food webs is beginning to provide crucial information that is needed in trying to relate population dynamic theory to questions of community structure. Memmott and Godfray (1993) recognise three different types of host–parasitoid webs:

(1) *connectance webs* that just record which hosts are attacked by which parasitoids (e.g. Rejmanek and Stary 1979; Hopkins 1984; Whittaker 1984);
(2) *semi-quantitative webs* that include information on the relative abundance of parasitoids on different hosts (e.g. Askew 1961; Force 1974; Askew and Shaw 1979; Hawkins and Goeden 1984) and;
(3) *quantitative webs* in which the abundance of all host and parasitoid species are recorded (Memmott *et al.* 1994; Müller *et al.* 1999; Schonrogge *et al.* 1999).

The latter are by far the most useful of the three in allowing the importance of the links where parasitoids attack more than one host to be evaluated. In particular, they enable one to gauge the extent to which webs may be broken down into functional compartments, recognised as semi-discrete units connected to the community as a whole by relatively few and weak connections. Such units are therefore dynamically compartmentalised to some extent (Pimm and Lawton 1980), and are the most amenable to being modelled as separate entities.

The more that webs can be broken down into such functional units of relatively few species, the more it becomes feasible and relevant to examine how such communities are influenced by the population dynamics of their constituent species (Begon *et al.* 1997; Holt 1997a). Most studies to date have simply built on the basic, two-species, Lotka–Volterra or Nicholson–Bailey framework by adding one extra species to give systems of two natural enemies attacking one host or prey (May and Hassell 1981; Hogarth and Diamond 1984; Kakehashi *et al.* 1984; Hassell and May 1986; Godfray and Waage 1991; Briggs *et al.* 1993), two hosts or prey attacked by a common natural enemy (Roughgarden and Feldman 1975; Comins and Hassell 1976; Holt 1977; Comins and Hassell 1987; Holt and Lawton 1994), or a tritrophic community composed of a host, parasitoid and hyperparasitoid (Beddington and Hammond 1977; Hassell 1978, 1979; May and Hassell 1981; Wilson *et al.* 1997). In this chapter we examine the dynamics of these various three-species systems upon which host–parasitoid communities are built, before concluding with a more elaborate community of five species—two host species, both attacked by their own specialist parasitoid species and also by a shared species.

6.2 One parasitoid attacking two hosts

Largely inspired by Paine's (1966, 1974) classic study on the role of the starfish, *Pisaster*, in promoting the coexistence of its competing prey, much of the early work on the dynamics of two prey species sharing a common natural enemy species sought to deduce how the shared natural enemy may influence the coexistence of prey species competing for the same resource (e.g. Cramer and May 1972; May 1974b; van Valen 1974; Murdoch and Oaten 1975; Rough-garden and Feldman 1975; Comins and Hassell 1976; Fujii 1977; Hassell 1979). More recently, the emphasis has changed towards examining the effects of a natural enemy shared between *non-competing* prey, and how this may give rise to 'apparent competition' (see below) (Holt 1977, 1984; Jeffries and Lawton 1984; Holt and Lawton 1993; Holt *et al.* 1994).

This section examines both kinds of interaction, starting with the more complete case of a shared natural enemy (= parasitoid species) attacking two hosts that compete, both intra- and interspecifically, for the same resource. The motivation here is to examine how natural enemies can alter the well-known criteria for coexistence and extinction in classical competition models. We then turn to the simpler interaction of two non-competing hosts linked only by the shared parasitoid species. This exposes the fundamental effects of a shared natural enemy in a more direct way, free of the additional effects of competition between the hosts for limiting resources.

6.2.1 Directly competing hosts

Let us consider the following scenario. Two competing insect herbivore species (N_1 and N_2) are attacked by a common parasitoid species (P) to give:

$$N_{i,t+1} = N_{i,t} \exp\left[r_i - \frac{r_i}{K_i}(N_{i,t} + a_i N_{2/i,t})\right] f_i(P_t), i = 1, 2$$

$$P_{t+1} = N_{1,t}[1 - f_1(P_t)] + N_{2,t}[1 - f_2(P_t)]$$

(6.1)

where:

$$f_i(P_t) = \left[1 + \frac{a_i P_t}{k_i}\right]^{-k_i}, i = 1, 2.$$

(6.2)

Parasitism is defined by the usual negative binomial model where a_i is the per capita searching efficiency and k_i is the degree of aggregation of attacks amongst hosts (Nicholson–Bailey when $k_i \rightarrow \infty$). In the absence of parasitism, we have a simple, discrete-generation, two-species competition model with host intrinsic rates of increase, r_i, carrying capacities, K_i, and the usual competition coefficients, a_i. The zero-growth isoclines for the two host species are linear, as in the comparable Lotka–Volterra model, and coexistence is thus feasible only if there is some niche separation ($a_1 a_2 < 1$). Coexistence is impossible if there is complete niche overlap ($a_1 a_2 \geq 1$) because the destabilising effects of inter-specific competition always outweigh the intraspecific density dependence. The primary difference between this discrete model and its continuous, Lotka–Volterra counterpart is in the ways that the populations can behave around their equilibria. Instead of always approaching the equilibrium mono-tonically, the inherent time delays of having a one-generation interval between cause and effect now also permit oscillatory damping, stable limit cycles and chaotic fluctuations (Hassell and Comins 1975).

Let us now introduce the parasitoid and consider the ways in which it can alter the conditions for coexistence of the hosts. There are three main effects.

Equivalent parasitism. If the parasitoids treat both host species identically (i.e. $a_1 = a_2$ and $k_1 = k_2$) and if the hosts have equal rates of increase ($r_1 = r_2$), then the natural enemy has no effect on the conditions for competitive co-existence (van Valen 1974; May 1977a), although it can alter the local stability properties of the coexisting hosts (Comins and Hassell 1976).

Preference. If, as is more likely, the parasitoids are more effective against one host than the other (i.e. $a_1 \neq a_2$ and/or $k_1 \neq k_2$), perhaps due to differential search rates, differing abilities of the hosts to escape and so on, then the host isoclines will be lowered by differential amounts. This makes it possible for the parasitoid to maintain a three-species equilibrium where the two host species alone could not coexist, provided that the interspecific competition is not too strong ($a_1 a_2 < 1$). This effect, however, cannot solely be interpreted in terms of differential parasitism. Exactly the same effect can be achieved with no parasitoid preference, but unequal prey rates of increase ($r_1 \neq r_2$).

Switching. The host preference above is constant, irrespective of the relative abundance of the different host species. In contrast to this, 'switching' allows the proportion of a particular host attacked to change from less than expected to greater than expected as the relative abundance of that host increases (Ivlev 1961; Murdoch 1969; Lawton *et al.* 1974; Cornell 1976; Cock 1978). In short, the most abundant host or prey type at any time suffers the greatest per cent mortality, as shown in Fig. 6.1

Such switching can have a profound effect on the dynamics of competing host or prey species (Elton 1927; Murdoch and Oaten 1975). Consider eqn (6.2) with switching introduced in a phenomenological way:

$$f_i(P_t) = \left[1 + \frac{(1 + E)a_i P_t}{k_i}\right]^{-k_i}, i = 1, 2 \tag{6.3}$$

where $E = s(N_1 - N_2)/(N_1 + N_2)$ and s is a constant expressing the degree of switching (Comins and Hassell 1976) (see Krivan 1996 for more sophisticated

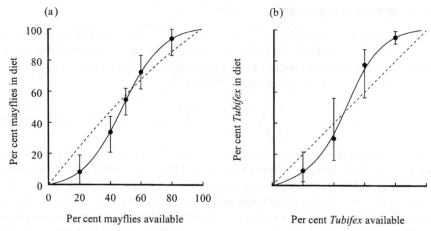

Fig. 6.1 Examples of switching in laboratory systems (means and ranges shown in both cases). (a) The per cent of mayfly (*Cloen dipterum*) larvae eaten by the predatory bug, *Notonecta glauca*, in relation to the relative abundance of mayflies and an alternative prey, *Asellus aquaticus*. (From Lawton *et al.* 1974.) (b) The per cent of *Tubifex* worms eaten by guppies in relation to the relative abundance of *Tubifex* and adult fruit flies, *Drosophila*. (From Murdoch *et al.* 1975.)

switching functions). Like the preference above, switching of sufficient magnitude can create a potentially stable, three-species equilibrium where none existed before when $a_1 a_2 < 1$. But it is a stronger effect that bends the host isoclines, and in this way can enable two (or more) host species showing complete niche overlap ($a_1 a_2 \geq 1$) to coexist (Roughgarden and Feldman 1975; Comins and Hassell 1976; May 1977*b*; Teramoto *et al.* 1979).

If switching is a widespread property of natural systems, it could clearly be an important process shaping community structure. But after considerable interest in the 1970s in demonstrating switching experimentally and in exploring its dynamical impact, there has been relatively little recent interest in the process. This is largely because it has been very difficult to demonstrate unambiguously in the field (e.g. Murdoch *et al.* 1984*b*) and also because interest has moved to other behavioural responses that can result in density-dependent changes in the frequency of attack (Sutherland 1983, 1996). For example, we shall see in the next section how a switching effect can readily arise from straightforward processes of natural enemy aggregation. A switching effect can also be obtained in other ways. For example, it has been shown that some parasitoids when they emerge as adults carefully examine the remains of the host, and use any volatile chemicals they detect as cues when searching for hosts. Hastings and Godfray (1999) incorporated this behaviour into a population dynamic model, and found that it could promote the coexistence of two hosts with a shared parasitoid species.

6.2.2 Apparent competition

Disentangling the precise contributions of intra- and interspecific competition and parasitism in model (6.1) is not straightforward. In this section we analyse a simpler three-species system of two non-competing host species (N_1 and N_2) linked only by their shared parasitoid species P:

$$N_{i,t+1} = \lambda_i N_{i,t} f_i(P_t), i = 1, 2$$
$$P_{t+1} = N_{1,t}[1 - f_1(P_t)] + N_{2,t}[1 - f_2(P_t)]. \tag{6.4}$$

Although the hosts do not compete directly for resources, the parasitoids produce the same kind of reciprocal negative effects on each host's growth rate that one would expect to observe in cases of true interspecific competition. This arises as follows. If the abundance of one host species increases, parasitoid numbers as a result will also increase. This, in turn, will lead to greater levels of parasitism on the other host species, whose population levels will therefore be depressed—just the result one would expect from a classical manipulation experiment to detect interspecific competition if one were unaware of the shared natural enemy. Because of these parallels with interspecific competition, Holt (1977, 1984) christened this type of indirect interaction as 'apparent competition'. Subsequently, Holt and Kotler (1987) also distinguished between 'short-term apparent competition' involving the immediate effects of be-havioural responses of the natural enemy to an increase in one of its hosts or prey, and 'long-term apparent competition' involving the population effects over several generations.

The potential importance of apparent competition in selecting for 'enemy-free space' (Jeffries and Lawton 1984) and in structuring insect communities in general has been widely discussed (e.g. Lawton and Strong 1981; Freeland 1983; Jeffries and Lawton 1984; Lawton 1986; Godfray 1994; Holt and Lawton 1994; Berdegue *et al.* 1996; Godfray and Müller 1998; Müller and Godfray 1999). Within model (6.4) the effects of apparent competition are dramatic and inevitably lead to one of the host species going extinct. Indeed, in an important paper, Holt and Lawton (1993) formally show that, irrespective of the form of $f(P_t)$ in (6.4) (as long as it is independent of N_1 and N_2), the two host species can never coexist in a stable equilibrium with the parasitoid (except in the unlikely situation of the parasitoid having exactly the same equilibrium density when supported on either host species alone). This is an interesting result and harks back to Nicholson (1933) who suggested that for a host–parasitoid equilibrium to exist there must be at least as many parasitoid species as host species, and to MacArthur's (1968) general assertion that there cannot be more species than there are niches. Suppose that survival from parasitism is again given by the negative binomial expression in eqn (6.2), so that the individual pairwise host–parasitoid interactions would be stable if, and only if, $k_i < 1$ (see p. 19). Such parasitoid density dependence, however, is *never* sufficient to enable the three-species system to persist. One host species always survives at the expense of the other, which Holt and Lawton show to be the one that supports the highest parasitoid density at equilibrium or, equivalently, the one that can withstand the higher density of parasitoids.[2] Host dominance is thus promoted by low values of a_i, and k_i and high values of λ_i.

In short, one species of host may even feed on a completely different resource from a second species and thus never encounter it in the field, and yet still be responsible for the exclusion of the latter because of indirect interactions mediated via a shared natural enemy. Holt and Lawton (1993) argue that such apparent competition is particularly likely to occur in host–parasitoid systems compared to predator–prey systems in general, since parasitoids show a numerical response on the same time scale as their hosts, and are often known to reduce host densities well below their carrying densities, thereby diminishing any density-dependent resource limitation of the hosts (Beddington *et al.* 1978; Lawton and McNeill 1979). Another outcome of apparent competition is that parasitoid species that are able to attack several species of hosts may actually appear to be specialists, not because of physiological or behavioural constraints but purely because of their effect on host coexistence: what Holt and Lawton (1993) have called 'dynamic monophagy'.

The empirical evidence for the importance of apparent competition in structuring insect communities is still sparse and largely based on field studies showing raised levels of mortality on one host species attributed to the presence of another species. For example, Settle and Wilson (1990*a,b*) have argued that the reduced population levels of the grape leafhopper, *Erythroneura elegantula*, that occurred after the spread of the congeneric *E. variabilis*, was caused not by interspecific competition between the two, but from parasitism by an egg parasitoid (*Anagrus epos*) whose numbers had built up to high levels on *E. variabilis*. Evans and England (1996) have similarly shown that the rate at which alfalfa weevils (*Hypera postica*) are parasitised by the ichneumonid, *Bathyplectes curculionis*, is increased by the presence of pea aphids (*Acrythosiphon pisum*). This is because *B. curculionis*, although not parasitising aphids, are attracted by the aphids' honeydew. Schonrogge *et al.* (1996) have found evidence for the natural enemies of alien cynipid galls on oak increasing the rates of attack on native gall formers. The only concrete example of long-term apparent competition in operation comes from a laboratory study of two stored product moths (*Plodia interpunctella* and *Ephestia kuehniella*) attacked by a shared parasitoid, *Venturia canescens* (Bonsall and Hassell 1997, 1998). In replicated time series the separate pairwise interactions between the parasitoid and one of the hosts were always stable. Three-species interactions were then carried out within cages that completely separated the two host species but allowed the parasitoids to roam freely attacking both hosts (Fig. 6.2). These three-species systems never persisted, with *E. kuehniella* always going extinct (Fig. 6.3).

This instability is strikingly at odds with the frequent field observation of

Side view 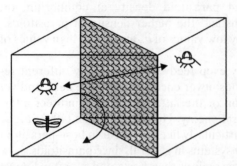 Plan view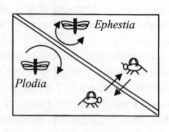

Fig. 6.2 Schematic design of experiments to demonstrate the effects of apparent competition in a laboratory system involving two species of stored product moths (*Plodia interpunctella* and *Ephestia kuehniella*) and their shared ichneumonid parasitoid, *Venturia canescens*. The diagonal barrier across the cage is made from nylon mesh which allows the parasitoids to pass freely through but not the hosts. (From Bonsall and Hassell 1997, 1998.)

coexisting hosts sharing common parasitoid species, and prompts the question of whether realistic biological processes have been omitted from model (6.4) that could override the effects of apparent competition. There are several candidate mechanisms which could have this effect. For instance, Holt and Lawton (1994) consider the possibilities that:

(1) hosts are resource-limited;
(2) a constant *number* of hosts escape parasitism in refuges (constant *proportion* refuges, however, always fail to allow coexistence); or
(3) switching may occur between the host species.

But they conclude that none of these are general or widespread enough mechanisms to have a major impact and that 'it is difficult for alternative host species to coexist when the sole regulatory factor impinging on them is a shared parasitoid'. Other mechanisms that could sometimes be important for coexistence in particular kinds of system or at certain spatial scales are: (1) that random environmental variation affects both host species in different ways so that the two host species achieve higher growth rates under different sets of environmental conditions (Chesson and Huntly 1989; Godfray and Müller 1998), although a stochastic version of the Holt and Lawton model gives the same criterion for extinction (Holt and Lawton 1993; Holt *et al.* 1994); or (2) that the interaction occurs at a metapopulation scale (see Chapter 7).

Of all these mechanisms, however, the one that may prove to be the most prevalent and important falls under the general heading of switching, simply because of the ease with which it may arise from the aggregative behaviour of parasitoids in a patchy environment (Comins and Hassell 1987; Bonsall and

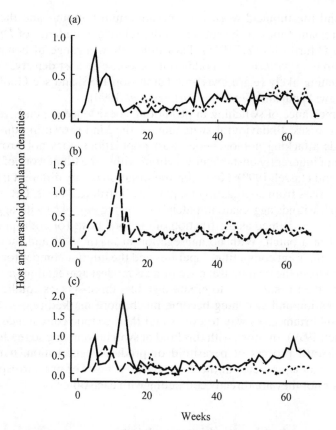

Fig. 6.3 Examples of time-series from the experiments illustrated in Fig. 6.2. (a) and (b) show the separate pairwise interactions, involving *Venturia canescens* (dotted line) and (a) *Plodia interpunctella* (solid line) and (b) *Ephestia kuehniella* (broken line). Combining the three species within the same system (c) leads to the rapid extinction of *Ephestia kuehniella*. (After Bonsall and Hassell 1997, 1998.)

Hassell 1999). The mechanisms underpinning switching relationships are varied. Since the original idea of a 'specific searching image' (Tinbergen 1960) that develops and wanes as particular prey increase and decline in relative abundance, the emphasis has changed to mechanisms in which natural enemies forage non-randomly in an heterogeneous habitat, allocating time to different areas depending on the relative abundance and distribution of the different host or prey species (Royama 1970). Consider, for example, a hypothetical situation in which parasitoids attack two separate host species, as in eqn (6.4), but in addition tend to aggregate on whichever host species is the most abundant at the time. This is a mechanism for producing a switching effect, in much the same way as in the example shown in Fig. 6.1(b), where the two prey types occupied quite different habitats—the *Drosophila* floating on the water surface and the tubificid worms on the aquarium bottom—and the guppies spent increasing times at the surface as the relative abundance of *Drosophila* increased (Murdoch *et al.* 1975). Parasitoids show a range of behaviours in which searching patterns are modified in response to host density, that make such switching likely (e.g. Lewis and Tumlinson 1988, and see Godfray 1994 for a review; Papaj and Vet 1990; Turlings *et al.* 1990).

The importance of switching in promoting coexistence depends very much on how it arises. Behavioural switching, of the kind shown in eqn (6.3), by parasitoids attacking homogeneous host populations does not promote co-existence of inherently unstable interactions within the framework of eqn (6.4) (Bonsall and Hassell 1999). However, the situation is very different if a switching effect arises from aggregation on patchily distributed hosts. Let us assume that the parasitoids aggregate in patches following eqn (4.6) with aggregative index, μ, and α_i defined as the fraction of the total hosts (of both species) that are found in a patch. With no aggregation at all ($\mu = 0$) and therefore no switching, apparent competition operates and the inferior competitor is always eliminated from the system. But once there is modest aggregation ($\mu = 0.5$) the switching effect is sufficient to create a stable, three-species equilibrium. As the aggregation and switching become much more marked (e.g. $\mu > 4$), the stable equilibrium gives way to cycles and then chaotic coexistence (Bonsall and Hassell 1999). In short, with this kind of switching arising so readily from a straightforward process of parasitoid or predator aggregation, it may well prove to be a major factor in the coexistence of hosts or prey, irrespective of whether or not they are directly competing with each other.

6.3 Two parasitoids attacking one host

For the inverted web of two parasitoid species attacking a single host species, the criteria for coexistence turns out to be quite different from those discussed above. As in all cases of direct competition, persistence requires that intra-specific effects must outweigh interspecific ones so that each parasitoid species has the advantage when it is rare. For competing parasitoids attacking a common host this can occur in several ways. There may, for example, be a colonisation–competition trade-off in a patchy environment, such that one species of parasitoid is inferior in exploiting patches but is a superior coloniser of patches (Lei and Hanski 1998). It therefore persists by virtue of its ability to find patches not yet discovered by the better competitor (see Chapter 7). Or, each parasitoid species may be favoured under certain conditions, perhaps caused by environmental variation or by fluctuations in host population densities that arise as part of the dynamics of the host–parasitoid interaction (Armstrong and McGehee 1980; Briggs 1993; Briggs *et al.* 1993). In this section we concentrate on two further mechanisms: (1) where there is a different kind of trade-off between the searching abilities of the adult parasitoids and their competitive abilities as larvae within hosts (Pschorn-Walcher and Zwölfer 1968; Zwölfer 1971); and (2) where the distributions of attacks by the parasitoid species are aggregated, either density dependently or density independently, thus once again serving to reduce the interspecific contacts.

We begin with the following scenario. A host species is attacked by two species of parasitoids, P and Q, each of which aggregates its attacks on hosts independently of host density and independently of the other parasitoid. This gives:

$$N_{t+1} = \lambda N_t f_1(P_t) f_2(Q_t)$$
$$P_{t+1} = N_t [1 - f_1(P_t)] \tag{6.5}$$
$$Q_{t+1} = N_t f_1(P_t) [1 - f_2(Q_t)]$$

where the parasitism survival functions, f_1 and f_2, come from the zero terms of the negative binomial (see eqn (6.2)). Model (6.5) makes precise assumptions about the timing of attacks by P and Q within the host's life cycle. It applies to cases where P acts first, followed by Q acting later on those hosts that escaped parasitism by P. Competition between P and Q for hosts is therefore indirect and mediated via the adults searching for a common host. Alternatively, and exactly equivalently, it also applies to cases where both P and Q act on the same host stage at the same time, but the larvae of P always outcompete those of Q should multiparasitism occur (e.g. Mackauer 1990). Dynamically, the two outcomes are identical. Only when the outcome of multiparasitism depends on the order of arrival within the host (Force 1970; Anderson *et al.* 1977; Laraichi 1978) is a rather different model structure required (Hogarth and Diamond 1984; Hochberg *et al.* 1990).

The conditions for coexistence in model (6.5) have been analysed by May and Hassell (1981), who drew the following conclusions.

1. A stable three-species equilibrium is most likely if both individual host–parasitoid links are stabilising; in other words, if k_1 and $k_2 < 1$ (see Fig. 6.4(a) for an example). In this case, the possibilities for stable coexistence are somewhat greater if species Q has the higher searching efficiency ($a_2 > a_1$). This is not surprising since Q has to overcome the disadvantage of having a smaller pool of healthy hosts to attack (if P and Q act in sequence), or of having competitively inferior larvae (if P and Q attack the same host stage). Such 'counterbalanced competition' (the species with the higher attack rate is the inferior larval competitor, and vice versa) has been suggested in several host–parasitoid systems and invoked as a general feature amongst coexisting parasitoids (Pschorn-Walcher and Zwölfer 1968; Zwölfer 1971; Schröder 1974).

2. Possibilities of coexistence are greatly reduced if only one parasitoid species contributes strongly to stability (e.g. $k_1 < 1$ and $k_2 \gg 1$). Indeed, it is not possible to have only one parasitoid species contributing to stability ($k < 1$) while the other species has the greater searching efficiency and hence the greater effect on depressing the host equilibrium level.

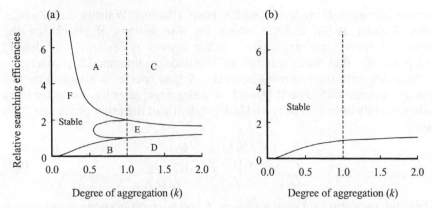

Fig. 6.4 Local stability boundaries for two kinds of two-parasitoid–one-host models. (a) Boundaries for two competing species of parasitoids attacking a single host species as in model (6.5), where f_1 and f_2 are defined by the negative binomial model and the host rate of increase, $\lambda = 2$. Boundaries are displayed in terms of relative searching efficiencies of the two parasitoids (a_2/a_1) against the degree of aggregated attacks, which in this case are equal for both parasitoid species ($k_1 = k_2 = k$). The three-species model is stable in region F and the dynamics in the other regions, A to D, are described in the text. (b) As in (a), but now the stability boundaries are for the host–parasitoid–hyperparasitoid model (6.7). Notice the much larger region in which the two parasitoids coexist. (From Hassell 1978.)

3. Three-species coexistence cannot occur if both P and Q attack randomly (unless there is additional host density dependence (see item 5)). Indeed, even an unstable three-species equilibrium can no longer exist for most parameter combinations (Nicholson 1933; Nicholson and Bailey 1935).

4. Outside the stable region shown in Fig. 6.4(a) a variety of different dynamics occurs. In regions A and C, species P always becomes extinct, leaving Q in a two-species interaction which is stable in A ($k < 1$) and unstable in C ($k > 1$). Thus the combination of the parasitoid density dependence, k, not being great enough and species Q having much the higher searching efficiency leads to the competitive replacement of parasitoid P. Regions B and D are the same as A and C, except that it is species Q that always goes extinct. A good example of parasitoid extinction apparently caused by competing species that can achieve higher levels of parasitism (presumably due to higher searching efficiencies) is shown in Fig. 6.5. Another classic example of successive replacements of introduced parasitoid species is that of three species of aphelinids introduced to control the citrus red scale (*Aonidiella aurantii*) in California (DeBach 1966; DeBach *et al.* 1971). This is now also interpreted in terms of the most efficient parasitoids replacing the previously introduced and competitively inferior species (Luck and Podoler 1985; Luck 1990). Finally, in

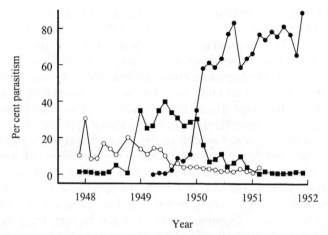

Fig. 6.5 Successive introductions and replacements of three species of parasitoids (*Opius* spp.) introduced into Hawaii for the biological control of the fruit-fly *Dacus dorsalis*. Hollow circles, *Opius longicaudatus*; solid squares, *Opius vandenboschi*; solid circles, *Opius oophilus*. (Data from Bess *et al.* 1961; after Varley *et al.* 1973.)

Region E (Fig. 6.4(a)) a three-species equilibrium exists, but the interaction is always unstable (cf. Fig. 6.7 below).

5. Additional density dependence acting on the host population can strongly promote stable coexistence as long as the parasitoids do not depress the host population too far below its carrying capacity (see Chapter 3, p. 44).

   In short, and unlike the two-hosts–one-parasitoid interactions described in the previous section, stabilising parasitism of the form of $f(P_t)$ can now readily promote coexistence. In an interesting extension to this work, Klopfer and Ives (1997) consider the impact of different kinds of parasitoid aggregation in relation to host density. They confirm, as in Chapter 4, that aggregation independent of host density (HDI) promotes coexistence, and that density-dependent aggregation (HDD) has a similar, albeit somewhat weaker, effect.

   May and Hassell (1981) also examined the extent to which the host equilibrium is depressed when one or both parasitoid species are present. Whether or not the addition of a second parasitoid species further reduces the host equilibrium below the level achieved by one parasitoid species on its own depends on the searching parameters, a and k. But the host equilibrium is never *higher* than that achieved by the more effective of the two parasitoids on its own. This conclusion is relevant to the long-standing debate on the relative merits of multiple versus single introductions of parasitoid species in biological-control programmes. Some have favoured finding the most efficient parasitoid species and only introducing this as the best way of minimising pest densities (e.g. Pemberton and Willard 1918; Turnbull and Charnt 1961; Watt 1965; Turnbull 1967). Others, however, believe that establishing additional parasitoid species will at best further reduce host densities and at worst do no harm (Smith 1929; Thompson 1939; Doutt 1961; van den Bosch 1968; Hassell and Varley 1969; Huffaker *et al.* 1971; May and Hassell 1981; Kakehashi *et al.* 1984).

The conclusion, that the introduction of a second parasitoid species will never have the adverse effect of creating host equilibria *above* those achievable with the most efficient parasitoid on its own, has been questioned by Kakehashi *et al.* (1984). They show that this result depends on the two parasitoid species having completely independent, negative binomial distributions of attacks. Instead of this they explored models of the form of eqns (6.5), except for the inclusion of different degrees of covariance in the probability distributions of hosts being attacked by the two parasitoid species. Thus, in the case of parasitoids P and Q responding to host cues in the same way, it is likely that the distribution of their attacks will be highly correlated, and it is then reasonable to replace the separate negative binomial terms in eqns (6.5) by a single negative polynomial with a common 'clumping' parameter, k:

$$f(P_t, Q_t) = \left[1 + \frac{a_1 P_t}{k} + \frac{a_2 Q_t}{k}\right]^{-k}. \tag{6.6}$$

The stability properties for three-species coexistence are little altered, but the best way of achieving the lowest host equilibria can be rather different. In particular, it is *always* better to introduce the most efficient parasitoid on its own rather than both parasitoids species together.

The real world is likely to lie somewhere between these two extremes of parasitoids having attack distributions that are either quite independent of each other, or covarying because they respond to the host's distribution and availability in exactly the same way. This intermediate stance blurs the conclusions; which is the most effective biological-control strategy now depends on knowing the actual parameter values of the parasitoids.

6.3.1 Age-structured interactions

A detailed stage-structured treatment of the one-host–two-parasitoid system has been given by Briggs (1993), and this provides a good comparison of the properties of two models of quite different complexity. While model (6.5) above (henceforth the 'M and H model') has some implicit stage-structure in assuming a sequence of parasitoid attacks during the hosts' life cycle, Briggs' model follows that of Murdoch *et al.* (1987) for modelling stage-structured populations. It deals explicitly with host eggs, larvae and adults, and immature and adult stages of the two parasitoid species. Parasitoid P only attacks host eggs while parasitoid Q only attacks host larvae. Two versions were considered: (1) where Q successfully parasitises only those larvae that escaped parasitism from P (e.g. Jones *et al.* 1993); and (2) where Q attacks all larvae, irrespective of whether or not they are already parasitised by P, and always survives at the expense of P should multiparasitism occur (e.g. Force 1974). In both cases, the only mechanism promoting stability is the presence of the invulnerable adult stage acting as a host refuge (see Chapter 5). This is in contrast to the M and H model where stability is due to the density dependence arising from small values of the parameter k. Briggs drew a number of interesting conclusions from her models.

1. With version (1) (Q only attacks survivors from P) parasitoid Q can only invade an existing $N–P$ interaction if it has a much higher searching efficiency than P, much as for the M and H model. However, unlike the M and H model, three-species coexistence was impossible: either P or Q alone can persist, and Q can only persist if it has a much higher searching efficiency than P. In short, Q is unable to coexist with P, simply due to the disadvantage of being able to reproduce only on those hosts left unparasitised by the earlier acting parasitoid, P.

2. In version (2), however, conditions for Q are improved by its larvae being able to survive in all hosts irrespective of whether or not they have already been parasitised by P, and three-species coexistence can now readily occur. When this happens, the density of each host stage is intermediate between the density obtained when either P or Q are present on their own.

Therefore, the greatest reduction in host equilibrium is always obtained with only the most efficient species of parasitoid present, a conclusion that is much more in line with that of Kakehashi *et al.* (1984) above. The key to understanding whether or not the establishment of a second parasitoid species necessarily further reduces host equilibrium levels appears to lie in the degree of independence of the two parasitoids in their attack distributions on the host (i.e. on the degree of their niche overlap, as first suggested by Turnbull and Chant in 1961). The M and H model has independently searching parasitoids and the two species always reduce the host equilibrium further than can one alone. In the Briggs' model the two parasitoids do not have independently aggregated distributions of attack, and the more efficient parasitoid always depresses the host equilibrium more than the two species together. Finally, Kakehashi *et al.* (1984) explored the continuum from independently searching parasitoids (the M and H model) to those responding to hosts in exactly the same way (see eqn (6.6)), in which case the more efficient parasitoid again always depresses the host equilibrium more than both species together. These results therefore suggest that similarities in the searching behaviour of parasitoids attacking a common host species may hold the key to whether single or multiple introductions of parasitoids leads to the lowest host equilibria.

Stage-structured multiparasitoid models can reveal other interesting properties that are not apparent from the simpler models. For example, Godfray and Waage (1991), in modelling the interaction between the mango mealy bug, *Rastrococcous invadens*, and two of its encyrtid parasitoids in West Africa, predicted that one parasitoid (*Gyranusoidea* sp.) should be clearly more efficient than the other (*Anagyrus* sp.) in reducing the host equilibrium densities, and that adding *Anagyrus* to a system already containing *Gyranusoidea* should lead to little improvement in the reduction of host equilibrium. Another example shows the value of modelling in explaining major features of biological-control programmes. Murdoch *et al.* (1966a) developed a stage-structured model for the interaction of California red scale with two of its parasitoids, *Aphytis lingnanensis* and *A. melinus*. Their models can account for the observed rapid competitive displacement of *A. lingnanensis* by *A. melinus* and the subsequent improvement in biological control in terms of quite subtle differences in the life histories of the two parasitoid species. In particular, the fact that a female *A. melinus* lays female eggs in smaller scales gives her a competitive advantage over *A. lingnanensis*, which therefore has a lower lifetime searching efficiency.

6.4 Host–parasitoid–hyperparasitoid systems

Insect food webs typically include a large number of secondary or hyperparasitoids that seek out and feed upon the immature stages of other parasitoids (the primaries) within their host (Godfray 1994). Most of these are obligate hyperparasitoids, but several species are facultative hyperparasitoids which can develop either as primary parasitoids within a healthy host or as secondary parasitoids if they encounter a host that has already been parasitised (Muesbeck 1931; Askew 1961). Because they can utilise a host, irrespective of whether or not it has been attacked by the primary parasitoid, a host–parasitoid–facultative hyperparasitoid system will be similar to the interaction with two competing parasitoids described in Section 6.2.1. Indeed, if Q is the facultative species and makes no distinction between healthy hosts and those parasitised by P, the interaction is identical to model (6.5) except for interchanging P and Q (May and Hassell 1981).

The first model of a true host–parasitoid–hyperparasitoid interaction was given by Nicholson and Bailey (1935) simply by extending their basic host–parasitoid model to include a randomly attacking hyperparasitoid:

$$N_{t+1} = \lambda N_t f_1(P_t)$$
$$P_{t+1} = N_t [1 - f_1(P_t)] f_2(Q_t) \tag{6.7}$$
$$Q_{t+1} = [1 - f_1(P_t)][1 - f_2(Q_t)]$$

where f_1 and f_2 are given by the zero terms of two Poisson distributions. The addition of the hyperparasitoid to the basic Nicholson–Bailey model raises the host equilibrium level, but still leaves the interaction unstable with rapidly expanding oscillations. Subsequently, Beddington and Hammond (1977) included hosts with a density-dependent, rather than constant, rate of increase [$\lambda = \exp(-gN_t)$] and concluded that:

(1) the establishment of a hyperparasitoid always raises the host equilibrium;

(2) the inclusion of host density dependence permits the three-species interaction to be stable, although the range of parameter values permitting this is small compared to the corresponding host–parasitoid system; and

(3) a stable, three-species system is most likely if the hyperparasitoid has a higher searching efficiency than the primary parasitoid.

The same model framework (6.7) was used by May and Hassell (1981) in which the host rate of increase was either constant or density-dependent, and the parasitism functions, f_1 and f_2, were given by the negative binomial model. Because of the stabilising parasitism, the three-species system can now be stable without any host density dependence, and is always stable if the hyperparasitoid has a higher searching efficiency than the primary and if both species strongly aggregate their attacks (k_1 and $k_2 < 1$), as shown in Fig. 6.4(b). Some empirical support for hyperparasitoids having higher searching efficiencies is given by Hassell (1979) and Kfir *et al.* (1976). Furthermore, for comparable parameter values, the area of stable parameter space is greater than for the equivalent interaction with two competing parasitoids (Fig. 6.4(a)). Coexistence with the hyperparasitoid having a *lower* searching efficiency is only possible if the aggregation of attacks for both species is more extreme. In all cases, the host equilibrium is higher with the hyperparasitoid present than with the primary parasitoid alone.

These models therefore suggest that hyperparasitoids are easier to establish than a parasitoid competing at the same trophic level, and that once established the consequences are always in one direction: the host equilibrium is raised.

6.5 Host–parasitoid–pathogen interactions

Interesting dynamics can also arise from a three-species system in which the host is attacked by a pathogen as well as a parasitoid species. Detailed laboratory experiments on this type of system have been carried out by Begon *et al.* (1997), using the flour moth, *Plodia interpunctella*, as the host, the ichneumonid *Venturia canescens* as the parasitoid and a granulosis virus (PiGV) as a pathogen of *Plodia*. Unlike the separate pairwise interactions that they set up, in which persistence was the rule with a marked tendency for single-generation cycles, the three-species systems failed to coexist in 12 out of 14 replicates (mean time to extinction of 286 days) and showed more complex dynamics with longer period cycles.

A theoretical treatment of a similar system, differing mainly in assuming that the host and parasitoid have discrete, synchronised generations, has been discussed by Hochberg *et al.* (1990). The pathogen is directly transmitted between hosts by the free-living infective stage that is continuously produced and released into the environment on the death of an infected host. The precise model is given by:

$$N_{t+1} = \lambda N_t (1 - I_t) f(P_t)$$

$$W_{t+1} = \frac{g\omega}{\beta} \ln\left(\frac{1}{1 - I_t}\right)$$

$$P_{t+1} = N_t (1 - \phi I_t)[1 - f(P_t)] \tag{6.8}$$

$$1 - I_t = \exp\left[-\frac{1}{N_T} (N_t I_t\{f(P_t) + \phi[1 - f(P_t)]\}) + \left(\frac{1}{\theta - q} W_t\right)\right]$$

where N_t, W_t and P_t are the densities of hosts (infected and uninfected), of infective stages and of parasitoids, respectively, and I_t is the fraction of the host population killed by the pathogen, all in generation t. The parameter λ is the host rate of increase and $f(P_t)$ is the host survival from parasitism as in eqn (6.2). The parameter β is the transmission coefficient for the rate of infection of susceptible hosts, ω is the rate at which the free-living infective stages of the pathogen are transferred to a reservoir of long-lived infective states, g is the fraction of the reservoir of infective stages that becomes accessible to the next generation of hosts and ϕ measures the dominance of the pathogen over the parasitoid when they co-occur with the same host individual. Thus the parasitoid larva outcompetes the pathogen when $\phi = 0$ and vice versa when $\phi = 1$. Finally, N_T is the critical density of susceptible hosts required to initiate an epidemic of the pathogen in the absence of long-lived stages and for $\phi = 1$.

The model combines the continuous production of infective stages and the continuous mortality of infected hosts within a single generation, with the discrete, generation-to-generation processes associated with the host and parasitoid. It is not surprising, therefore, that the dynamics are complex. Briefly, if the parasitoid searches randomly (Nicholson–Bailey), three-species coexistence requires that there is counterbalancing between the relative searching or transmission abilities (a and β) and the outcome of competition within any jointly infected host individuals (ϕ). If, however, the parasitoid attack is more clumped than random (i.e. k is small), this trade-off is blurred and coexistence may arise even when the parasitoid is both the better intrinsic and extrinsic competitor. Depending on the parameters, this coexistence may be associated with constant, cyclic or chaotic populations (Fig. 6.6). It is unclear why the experiments of Begon *et al.* (1997) failed to show any of this coexistence. One obvious difference between the two is the overlapping of generations in the real system in contrast to the discrete generations in the model. But, if anything, this should have made the experimental system more stable than the model. Perhaps there was an increased risk of extinction arising from the relatively small scale of the laboratory system.

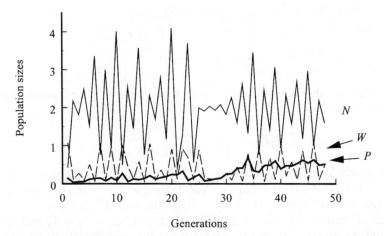

Fig. 6.6 Changes in host (N), pathogen (W) and parasitoid (P) numbers with time from model (6.8), showing chaotic dynamics from a system in which the separate pairwise interactions are both unstable. Parameters: $\lambda = 5$, $g = 0$, $\omega = 0.01$, $\mu = 0.01$, $\beta = 0.05$, $a = 1$, $k = \infty$, $\phi = 0.5$, $d = 0$. (From Hassell and Anderson 1989.)

6.6 More complex webs

The different kinds of three-species systems outlined in the previous sections can be linked together to make more complex food webs, particularly by the addition of generalist and shared oligophagous natural enemies. Understanding the dynamics of these more complex communities therefore requires not only a basic understanding of the different two- and three-species modules, but also knowing how easily different trophic configurations can be 'invaded' by additional species and then whether or not these new, more species-rich, communities can persist. If the whole community fails to persist, then we also need to know where the breaks in the food web are most likely to occur and which simpler trophic configurations are the most robust.

In this section this approach is illustrated using a simple, five-species community of hosts and parasitoids, in which two hosts (N_1 and N_2), that do not interact directly, are attacked by three species of parasitoid—two specialists (P_1 and P_2), one on each host, and one shared species (Q) attacking both hosts. The model is a straightforward extension of the three-species models above, with the additional assumption that the larvae of Q are competitively superior to those of either P_1 or P_2 should multiparasitism occur. This is justified on the basis that parasitoids with a broader host range are more commonly ecto-parasitoids that tend to kill or paralyse their hosts at oviposition, in contrast to specialist parasitoids which tend to be endoparasitic (Askew and Shaw 1986; Sheehan and Hawkins 1991; Godfray 1994). We therefore have:

$$N_{i,t+1} = \lambda_{i,}N_{i,t}f(P_{i,t})f(Q_t), i = 1, 2$$

$$P_{i,t+1} = N_{i,t}f(Q_t)[1 - f(P_{i,t})], i = 1, 2 \qquad (6.9)$$

$$Q_{t+1} = \sum_{1}^{2} N_{i,t}[1 - f(Q_t)].$$

Parasitism by the two specialist species is defined by the usual negative binomial function

$$f(P_{i,t}) = \left[1 + \frac{a_{p,i}P_i}{k_i}\right]^{-k_i}, i = 1, 2 \qquad (6.10$$

and similarly for species Q. Full details of this model and its analysis are given in Wilson *et al.* (1996). Here we just look at the simple case of the two hosts having the same rate of increase ($\lambda_1 = \lambda_2 = \lambda$), all the parasitoids showing the same degree of density dependence in their attack rate ($k_1 = k_2 = k$) and the two specialist parasitoids, P_1 and P_2, having the same searching efficiencies, a_P. All these assumptions can be relaxed without overturning the general conclusions. First let us consider whether or not an absent parasitoid can invade the system, which can be determined from its growth rate when its population size is very small. Whichever parasitoid is invading, Q or P, the invasion criteria depend on the searching efficiencies of the invader being sufficiently high relative to the resident species (i.e. a_P versus a_Q and vice versa) (Wilson *et al.* 1996).

The full invasion boundaries are plotted in Fig. 6.7 as a function of the density-dependent parasitoid attack rate, k, and the relative searching efficiencies, a_P/a_Q, and show which combinations of species can persist. First we note that no persistent communities can occur unless there is a threshold level of parasitoid density dependence ($k < 1$), in which case four separate regions of parameter space are distinguished (A to D). In region A the specialists have too high a searching efficiency for the parasitoid Q to invade (note that the upper boundary is identical to that in Fig. 6.4(a)). In region B, however, parasitoid Q has a higher relative searching efficiency which, together with its competitively superior larvae, is enough to preclude the specialists so that a community with only the two hosts and parasitoid Q occurs. Region C is the only area in which a stable five-species community is possible, while in region C* all species can persist but limit cycles or chaotic dynamics are observed. Finally, in region D more complex dynamic behaviour occurs which Wilson *et al.* (1996) have dubbed a region of 'complex dynamic instability'. In this interesting region all three parasitoid species can invade if not already present, yet the full five-species community shows large amplitude fluctuations in density leading inevitably to the extinction of the two specialists. Once extinct, however, either or both specialists are able to reinvade again, leading to a four- or five-species community that will then break down again with the loss of the specialists. The community observed at any one time therefore depends critically on the speed with which the five-species community collapses, and the frequency with which the specialists reinvade.

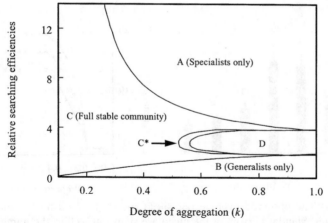

Fig. 6.7 Invasion boundaries for the five-species model (6.9) in terms of the relative searching efficiencies (a_P/a_Q) of the specialists P and the shared parasitoid Q (labelled the 'generalist'), compared with the degree of density dependence in the parasitoid attack (k in eqn (6.10)). All parasitoids have the same value of k, both specialists have the same searching efficiency, $a_P = 0.3$, and both hosts have the same rate of increase, $\lambda = 2$. The different regions are explained in the text. (From Wilson *et al.* 1996.)

This community shows an impressive range of possible dynamical behaviour. Despite its simplicity, it is also more or less at the limit of complexity for developing tractable models in which each population is separately formulated. A rather different approach to understanding and modelling the dynamics of host–parasitoid communities has been taken by Hochberg and Hawkins (1992, 1993) and Hawkins *et al.* (1993), who seek to explain their observed pattern of parasitoid species abundance in relation to the hosts' feeding niches (Fig. 6.8(a)). Hochberg and Hawkins assume that there is a gradation of increasing degrees of host refuge as one moves from external feeders to root feeders, and on this basis have developed a model in which parasitoid species diversity is compared with the proportion of hosts in any generation within a refuge. The model is made up of a discrete-generation host population that is initially attacked by 50 species of specialist parasitoids and 50 species of generalists. The hosts have a constant rate of increase and a specified fraction are protected from the specialists and generalists within a refuge. The attack rates of the natural enemies are given by type II functional responses with negative binomial distributions of attacks (see Chapter 2, Row F in Table 2.1). For a given set of parameters, the proportion of hosts in the refuge is varied and the model simulated for 1000 generations to determine how many specialists and generalists fail to persist in the system over this period. An example of their results is shown in Fig. 6.8(b); species diversity is low at both small and large refuge sizes giving the same 'humped' shape to the diversity curve as in Fig. 6.8(a). Hochberg and Hawkins (1992) explain this pattern in terms of a

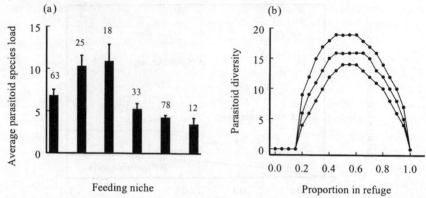

Fig. 6.8 (a) The observed average parasitoid species load on hosts with different feeding niches. Standard errors of the mean are shown as well as the number of host species in each category. From left to right, the different feeding niches are: external feeders, rollers/webbers, leaf-miners, gallers, borers and root feeders. (From Hochberg and Hawkins 1992.) (b) Numerical simulations showing the predicted change in parasitoid species diversity as the proportionate refuge for the hosts increases (further details of the model are given in the text). The three curves are for different degrees of aggregation of parasitoid attacks on hosts; the upper curve has the highest aggregation ($k = 0.25$), the middle curve is for $k = 0.5$ and the lower curve is for $k = 2$. (From Hochberg and Hawkins 1992.)

fundamental way that refuges from mortality can influence species diversity. However, as Godfray (1994) points out, the dome-shaped curve from the model can also be explained by the way that the model is set up and the way that refuges in the model affect host equilibrium abundance. At low refuge levels, the parasitoids cause high host mortality and host abundance is therefore low so that fewer species can be supported. At high refuge levels, there are relatively few hosts exposed to parasitoids and, again, fewer species can therefore be supported. Only at intermediate refuge levels can many species of parasitoids be maintained on abundant hosts. This view is further supported from Fig. 6.8(b) where the parasitoid diversities are increased by higher levels of parasitoid density dependence (smaller k), which in turn lead to the higher host equilibrium levels. More studies are needed to unravel the ecological basis for these intriguing patterns of community structure.

6.7 Summary

Our current understanding of host–parasitoid population dynamics comes largely from models of single-host–single-parasitoid interactions. Onto this model framework has been added additional host or natural enemy species to make a variety of three-species systems: two host species attacked by a common parasitoid species; two parasitoid species attacking a common host species; and host–parasitoid–hyperparasitoid and host–parasitoid–pathogen interactions. This chapter examines the dynamics of these interactions.

The ways that a shared natural enemy can affect the coexistence of two host species can be complex. If the hosts compete directly for resources, the natural enemy can alter the classical conditions for coexistence between the competing species. For example, both preference by the parasitoid for one host over the other and, to an even greater extent, 'switching' between hosts, can enhance the conditions for coexistence. In particular, switching is a mechanism that enables competing hosts to coexist even if they show complete niche overlap. If the hosts do not compete directly for resources, the same kinds of reciprocal negative effects on each other's growth rate, that one expects from interspecific competition, can still occur due to indirect effects mediated via the shared natural enemy species. This has been dubbed 'apparent competition'. Simple models with apparent competition between two hosts and a shared parasitoid are unstable with one host being driven to extinction. A variety of mechanisms exist that counteract this and enable the system to persist. Of these, the most important is likely to be the aggregation of parasitoids to overall host density which produces a stabilising, switching effect.

Adding an additional parasitoid species to a host–parasitoid interaction can also lead to a wide range of interesting dynamics. Depending on how stabilising are the individual host–parasitoid links (e.g. due to density-dependent parasitoid attack rates), the parasitoid may fail to invade, replace the other species, or all three species may coexist, in which case the host equilibrium level may be reduced or increased compared to its previous level. In general, such interactions with two parasitoid species attacking the same hosts tend to be more stable than the comparable two-host–one-parasitoid systems, provided that the individual host–parasitoid links are themselves stabilising (i.e. $k < 1$).

Adding a hyperparasitoid to a host–parasitoid interaction produces yet another possible three-species configuration. Coexistence is again enhanced by density dependence acting on the parasitoids, and also by the hyperparasitoid having a higher searching efficiency than the primary parasitoid species. In general, host–parasitoid–hyperparasitoid systems are more stable than a three-species system with two primary parasitoids attacking the same host species.

Host–parasitoid–pathogen interactions also show a wide range of dynamics depending on the demographic parameters and the competitive abilities of pathogen and parasitoid within the same attacked host. Where coexistence occurs, the populations may be constant, cyclic or chaotic depending on the parameters.

Finally, a five-species host–parasitoid system is examined involving two host species, each with its own specialist parasitoid, and an additional natural enemy shared between the two. No persistence occurs unless there is sufficient density dependence acting on the parasitoids ($k < 1$), in which case four different outcomes can be distinguished:

(1) a three-species system comprised of the two hosts and their shared natural enemy;
(2) two-species systems made up of each host with its specialist parasitoid;
(3) all five species coexisting; and
(4) a region of dynamic complexity in which species may invade and persist for some time before going extinct, followed by subsequent reinvasions.

Notes

1. Definitions of 'oligophagy' abound, all rather vague and in terms of parasitoids whose host range falls between specialists and polyphages. Here the term is used in a more dynamical way: species whose dynamics can neither be decoupled from their hosts, nor understood by considering one host species in isolation.
2. Note the equivalence with Tilman's (1982) model of differential resource utilisation in plants, in which the winner is the species that reduces resource supply to the lowest level.

7

Metapopulations of hosts and parasitoids

7.1 Introduction

The spatial interactions between hosts and parasitoids discussed in Chapter 4 assume complete mixing of the dispersing individuals of both populations at least once during each generation period. It is as if the individuals enter a 'global pool' and then redistribute themselves amongst the patches according to particular rules. This implies that individuals are able to disperse widely across their entire habitat which, in turn, restricts the spatial scale that is appropriate for such systems. A very different picture can emerge, however, if the habitat is on a much larger scale relative to the characteristic dispersal rates of the organisms (see Levin (1992) and Levin (1994) for a general treatment of the problem of pattern and scale in ecological interactions). These larger patches now contain entire local populations that are only partially linked by dispersal. Some degree of autonomous population dynamics can therefore easily occur over a period of generations, depending on the type and degree of dispersal from other local populations. Such an ensemble of partially independent local populations forms a 'metapopulation', a term originally coined by Levins (1968, 1969, 1970) (see below). Iwao (1971) describes just such a system. Populations of the phytophagous lady beetle, *Epilachna vigintioctomaculata*, have discrete annual generations and, because of the patchy distribution of suitable habitats, are divided into distinct subpopulations. These are interconnected by adult movement which, rather than involving complete mixing amongst all subpopulations, is limited to a small range. Iwao concludes that 'The numbers in the ... subpopulations tended to be stabilised through population interchange among them, density-dependent adult mortality ... and larval competition for food'. Nicholson and Bailey (1935) clearly also had such a scheme in mind in trying to reconcile the instability of their basic host–parasitoid models (Chapter 2) with the persistence that is widely observed in the field: 'the interacting animals exist in numerous disconnected small groups, within each of which interspecific oscillations follows its course independently of that in the other groups.' Similarly, Nicholson (1947) wrote that the 'cycle of increase in numbers, followed by ... extermination, proceeds independently in different parts of the occupied country; so at all times some groups are increasing and some decreasing in numbers ... Consequently when one considers a large tract of country, the abundance [of both host and parasitoid] ... remains more or less constant; whereas in any small area of the same country the fluctuation in numbers … may be violent.' As Hanski and Simberloff (1997) have pointed out, Andrewartha and Birch, despite their antagonism to almost all of Nicholson's views, at least shared common ground here in emphasising the frequency of local population extinction: 'spots that are occupied today may become vacant tomorrow and reoccupied next week or next year' (Andrewartha and Birch 1954).

Metapopulation models have developed in quite different ways (see reviews by Taylor 1988b; Kareiva 1990; Hanski and Simberloff 1997; Nee *et al.* 1997; Hanski 1999). In the original formulation of Levins (1968) space was implicit; the habitat was effectively divided into an infinite number of patches of equal size and quality, all of which are equally in contact with each other and are either empty, or occupied by the species in question. The models record the rate of change in the proportion of occupied patches as a function of colonisation and extinction parameters without any explicit consideration of local dynamics. The implicit assumption is that populations within patches go immediately to their carrying capacity, then remain there as long as the patch stays occupied. There is thus the implicit assumption of very strong, within-patch density dependence. There have been many developments based on this framework; for example, allowing stochastic colonisations and extinctions (e.g. Gurney and Nisbet 1978; Hanski 1991, 1997*a*), spatially correlated environmental stochasticity (e.g. Hanski 1996) and varying patch quality, population growth as well as extinction and colonisation rates (see review by Hanski 1997*a*). These kinds of metapopulation models have also been extended to describe two-species interactions (e.g. May 1994; Holt 1997b; Nee *et al.* 1997) and multispecies communities (Tilman 1994; Wennergren *et al.* 1995), looking at how trade-offs between extinction and colonisation can facilitate coexistence.

Another class of metapopulation models has taken a spatially explicit approach in defining habitats or patches within a grid-like arrangement so that each patch has specific neighbours. Some of these 'cellular' models also record only different states of habitat occupancy (e.g. Maynard Smith 1974; Wolfram 1984; Crawley and May 1987; Wilson *et al.* 1993; Durrett and Levin 1994; Nowak *et al.* 1994; Rand *et al.* 1995; Rand and Wilson 1995). Alternatively, local populations may be represented by continuous variables and have explicit dynamics (e.g. Reeve 1988; Hastings 1990; Taylor 1990; Comins *et al.* 1992; Solé and Valls 1992; Rohani and Miramontes 1995; Swinton and Anderson 1995; Wilson *et al.* 1997, 1998).

Four conditions are necessary for such metapopulations to persist (Hanski and Kuussaari 1995; Walde 1995):

(1) the local populations should breed in relatively discrete patches;
(2) no local population should be large enough and persist for long enough to constitute a 'mainland population';
(3) the local habitat patches should not be sufficiently isolated to prevent recolonisation; and

(4) the local population dynamics should be sufficiently asynchronous to make their simultaneous extinction highly unlikely.

There have been many recent attempts to detect metapopulation dynamics under natural conditions (see Harrison 1991; Schoener 1991; Harrison and Taylor 1997 for reviews). However, populations that unequivocally fall into this category are still relatively few. The clearest examples amongst arthropods come from butterfly populations (e.g. Thomas and Hanski 1977; Harrison *et al.* 1988; Hanski *et al.* 1995*b*; Gyllenberg and Hanski 1997; Hanski 1997*b*; Harrison and Taylor 1997; Nieminen and Hanski 1998; Saccheri *et al.* 1998; Hanski 1999); mites (e.g. Sabelis *et al.* 1991; Walde 1991, 1994, 1995), waterfleas (Hanski and Ranta 1983) and spiders (Schoener and Spiller 1987*a,b*; Spiller and Schoener 1990). There has also been an elegant laboratory demonstration of meta-population dynamics by Holyoak and Lawler (1996) working with a protist predator–prey interaction in a system of interconnected patches.

There are several excellent examples of how metapopulation models can be applied to real systems (e.g. Hanski 1994; Brodmann *et al.* 1997; Hanski 1997*b*; Hastings *et al.* 1997; Maton and Harrison 1997; Hanski 1999). Hanski has developed what he calls 'incidence function' metapopulation models. These differ from the Levins model in three ways that greatly facilitate the link between models and real metapopulations in the field:

(1) local populations are now confined to a finite rather than infinite number of local habitat patches;
(2) each of these has its own spatial co-ordinates; and
(3) local populations interact most strongly with their nearest neighbours, rather than interacting globally.

The model is therefore well suited for exploring the effects of area and the relative isolation of patches on the metapopulation dynamics of different systems. It has enabled Hanski and his colleagues to make a wide range of both fundamental and practical predictions of factors affecting colonisation and extinction for a number of butterfly populations that they have studied over several years (Hanski 1994; Hanski *et al.* 1995*a*, 1996; Wahlberg *et al.* 1996). Lei and Hanski (1997, 1998) have also applied an incidence function model in their metapopulation study of the butterfly, *Melitiaea cinxia*, and its specialist braconid parasitoid, *Cotesia melitaearum*. Their results indicate a classical metapopulation structure for the interaction 'with the incidence of the parasitoid increasing with increasing host population size, and with decreasing isolation' (Lei and Hanski 1997). Also, provided hyperparasitism is not too heavy, the *Cotesia* tend to drive the hosts to extinction in all but the smallest habitat patches; the parasitoids are then unable to maintain themselves in these and become extinct.

This chapter concentrates primarily on explicitly spatial metapopulations applied to host–parasitoid interactions. The key questions concern how population persistence is affected:

(1) by a metapopulation structure *per se*;
(2) by the physical configuration of the metapopulation (e.g. the number of patches and their precise spatial distribution); and
(3) by different kinds of stochasticity.

Since the amount of mixing between local populations is a major factor in determining the kinds of metapopulation dynamics observed, the chapter is broadly divided on the basis of whether dispersing individuals mix thoroughly over the entire habitat, or show more restricted dispersal around the area where they developed.

For the main part, the total environment is assumed to be subdivided into local habitats (= patches), that are arranged in a grid or lattice, each containing the foodplant for an herbivorous insect that is attacked by a specialist parasitoid. We shall assume that the insects have discrete generations and within each generation there are two distinct phases: (1) *a between-patch dispersal phase*, in which a proportion of adult hosts and parasitoids leave their natal patch to colonise other patches according to some dispersal rule that determines the degree of mixing within the metapopulation as a whole; and (2) *a within-patch parasitism and reproduction phase*, in which the local host population is parasitised and the surviving healthy hosts go on to reproduce. Reproduction can thus occur following, or prior to, dispersal, but not at the same time (this would require a different and detailed formulation of the model to keep track of those individuals that migrated and those that remained developing within the patch). Such models with discrete time and space, but continuous population state have been dubbed 'coupled map lattices' (Kaneko 1992, 1993; Solé *et al.* 1992; Solé and Valls 1992).

7.2 Global dispersal

We begin with a special case: the dispersing hosts and parasitoids spread with equal likelihood throughout the whole area. Such global dispersal means that space can usually be treated implicitly; the co-ordinates of each patch are irrelevant if each patch is colonised with the same probability. On the spatial scale of most metapopulations this complete mixing would be unlikely, but it does serve to emphasise an important limiting case whereby the metapopulation becomes the same as the single-patch model on which it is based. Thus, the Nicholson–Bailey model is *exactly* recovered from a lattice metapopulation in which the hosts within each patch have a constant rate of increase and suffer Nicholson–Bailey parasitism (constant searching efficiency and a linear functional response) and, at some different stage in their life cycles, *all* adult hosts and parasitoids disperse from their natal patch to colonise any other patch with equal probability (Reeve 1988; Comins *et al.* 1992; Wilson and Hassell 1997). Quite simply, the global mixing ensures that all patches are synchronised in their dynamics.

Parallel work has examined global dispersal in continuous-time meta-populations based on the Lotka–Volterra predator–prey model. Roos *et al.* (1998) considered an *n*-patch system in which prey dispersal was limited, while the predators dispersed globally to attack all patches equivalently. Under the conditions of no prey dispersal, exponential prey increase and a linear predator functional response, the homogeneous Lotka–Volterra model is exactly recovered. However, with logistic prey growth and a type II predator response, a variety of stable cyclic behaviour was observed, in some cases with interesting spatial patterns of prey abundance.

Lattice metapopulation models only become really distinctive once dispersal between patches is limited in some way. Following Reeve (1988, 1990), let us consider a host–parasitoid system, differing from the Nicholson–Bailey lattice model above only in:

(1) a fraction, μ_H of adult hosts and μ_P of parasitoids, disperse in each generation;
(2) that parasitism within patches is described by the negative binomial distribution (with clumping parameter k) rather than the Poisson; and
(3) environmental stochasticity is included in the form of patch-to-patch random variation of both parasitoid attack rates and the host rate of increase.

Thus, assuming that dispersal in each generation occurs in the time interval t to $t + h$, the number of hosts in patch i following dispersal is given by:

$$N_{i,\,t+h} = (1 - \mu_N)N_{i,\,t} + \left(\frac{1}{n}\sum_{j=1}^{n} \mu_N N_{j,\,t}\right) \tag{7.1}$$

where n is the number of patches. A similar equation applies to the parasitoids. Following this dispersal, host reproduction and parasitism occur in the usual way to give:

$$N_{i,\,t+1} = \lambda_i N_{i,\,t+h}\left(1 + \frac{a_i P_{i,\,t+h}}{k}\right)^{-k}$$

$$P_{i,\,t+1} = N_{i,\,t+h}\left[1 - \left(1 + \frac{a_i P_{i,\,t+h}}{k}\right)^{-k}\right] \tag{7.2}$$

where λ_i is the host rate of increase and a_i is the parasitoid searching efficiency of the P_i adult parasitoids, all per patch i. First we note that, in the absence of environmental variability, stable local populations ($k < 1$) lead to stable metapopulations (see also Allen 1975; Crowley 1981; Taylor 1988b; Hastings 1990; Taylor 1990). With $k > 1$, local populations are unstable, and this destabilises the whole metapopulation, except for a range of very low dispersal rates when sufficient variation in initial host and parasitoid densities can lead to long-term persistence of the whole system (Adler 1993; Taylor 1998). Introducing environmental stochasticity, however, can considerably broaden the conditions for persistence. Reeve illustrated this from simulations of his model in which λ_i and a_i were randomly distributed from patch to patch. Metapopulation persistence was increased, even when the local population dynamics were completely unstable ($k \rightarrow \infty$) (see also Adler 1993; Taylor 1998 for similar conclusions). The effect arises through the random variation causing asynchrony in the dynamics of the individual patches. Hence, patches in which either the host or parasitoid go extinct due to the unstable local dynamics can be recolonised by dispersal from other patches (a rescue effect) without the metapopulation as a whole becoming extinct. Anything that promotes this asynchrony—low dispersal rates (μ_H and μ_P) independent of population densities and high levels of variability in host and parasitoid growth rates—increases the tendency for metapopulation persistence.

A quite different form of random variation arises due to demographic stochasticity. The models above have local population sizes as continuous variables (i.e. fractional individuals are possible at very low densities) and are formulated as mean field approximations to the biological process of individual parasitoids finding hosts. But if the numbers of individuals within patches fall to very low levels, the process may not be well described by this mean approximation because of the effects of demographic stochasticity. In general, therefore, demographic stochasticity can have an important influence on dynamics whenever population sizes become very small. The effects of this in the models above depend on the dispersal rates. With complete global dispersal ($\mu_H = \mu_P = 1$), the Nicholson–Bailey lattice model is unaffected, except for demographic stochasticity tending to reduce overall parasitoid attack rates and therefore increasing population levels. But if only a fraction of hosts and parasitoids disperse (μ_H and $\mu_P < 1$) as assumed by Reeve above, the demographic stochasticity has the same effect as environmental stochasticity: the individual patches fluctuate out of phase with each other which, if sufficient, leads to the metapopulation as a whole persisting. The importance of demographic stochasticity to metapopulation dynamics is more fully discussed in the next section.

Global dispersal of individuals over an entire habitat region is unlikely at the scale of most real metapopulations; rather, individuals' movements will be concentrated around the area in which they developed as juveniles. The introduction of such restricted individual movements ('local dispersal') can have profound effects on metapopulation dynamics: it is an important cause of the asynchrony between local populations needed for metapopulation persistence, and it leads to travelling waves of population abundance as individuals spread within the metapopulation area.

7.3 Local dispersal

Let us consider the same scenario outlined above, except that dispersal is now restricted only to the immediately adjacent local populations. Thus, in each generation a certain fraction of adult hosts, μ_H, and adult female parasitoids, μ_P, leave the patch from which they emerged, while the remainder stay behind to complete their life cycle in their original patch. The dispersing individuals, however, rather than dispersing throughout the area, spread outwards to colonise equally the eight nearest neighbours of the patch from which they emerged. Longer range dispersal can therefore only occur through repetition of these single-patch movements over a number of generations (more complex dispersal rules are discussed in Rohani and Miramontes (1995) and Wilson *et al.* (1997), and see p. 144).

The model is defined in two parts. First, the equations for the dispersal stage in each patch are given by:

$$N'_{i,t} = (1 - \mu_N)N_{i,t} + \mu_N \bar{N}_{i,t}$$
$$P'_{i,t} = (1 - \mu_P)P_{i,t} + \mu_P \bar{P}_{i,t} \tag{7.3}$$

where $N_{i,t}$ and $P_{i,t}$ are the pre-dispersal host and parasitoid population densities in patch i at time t, $N'_{i,t}$ and $P'_{i,t}$ are the densities after dispersal, and $\bar{N}_{i,t}$ and $\bar{P}_{i,t}$ are the host and parasitoid populations averaged over the eight nearest neighbouring patches (Hassell *et al.* 1991a; Comins *et al.* 1992). Slightly different definitions are needed for patches along the boundary of the arena, depending on whether one assumes *cyclic*, *absorbing* or *reflective* boundary conditions. Cyclic boundaries have opposite edges of the arena effectively joined together, which is obviously unrealistic, but has the advantage that all patches are dynamically equivalent, with no edge effects. Absorbing boundaries remove individuals as they cross the boundary, which therefore imposes an additional mortality on the system. Reflective boundaries prevent individuals from crossing so that they remain in the edge patches. As long as the arenas are sufficiently large, the type of boundary condition assumed generally makes little difference, except that simulations with cyclic boundaries tend, not surprisingly, to produce more symmetrical spatial patterns of abundance.

The second part of the model defines the dynamics within a patch:

$$N_{i,t+1} = \lambda N'_{i,t} f(P'_{i,t})$$
$$P_{i,t+1} = N'_{i,t}[1 - f(P'_{i,t})] \tag{7.4}$$

where λ is the constant host rate of increase, and survival from parasitism is defined either by the negative binomial or Nicholson–Bailey models.

This introduction of explicit space with nearest neighbour movement has profound effects on dynamics (Hassell *et al.* 1991a, 1994; Comins *et al.* 1992; Comins and Hassell 1996). These are outlined in the following four sub-sections.

7.3.1 Stable local populations

As found by Reeve (1988), the system as a whole will be stable if the within-patch, local populations are also stable (i.e. $k < 1$). In other words, stable and persisting local populations prevent the 'turnover' of habitat patches that is a characteristic feature of a metapopulation. This result makes sense intuitively. Provided that the environment is uniform, all local populations at equilibrium will have the same density, and movement to and from local populations will be in balance. The same applies to comparable systems with single, or competing, species interactions (Hassell *et al.* 1995; Rohani *et al.* 1996). A number of factors will, of course, confound this simple conclusion. Most obviously, varying habitat 'quality' affecting the demographic rates of the local populations is bound to introduce different dynamics depending on the nature of the heterogeneity that is imposed.

7.3.2 Spatial patterns of abundance

The dynamics become markedly different, however, if we assume local population dynamics that are unstable (e.g. Nicholson–Bailey). Any metapopulation persistence that now occurs will result from the spatially distributed nature of the interaction, rather than from within-patch stability. The combination of restricted dispersal and 'boom and bust' local patch dynamics leads to a tendency for travelling waves of host and parasitoid abundance moving about the metapopulation arena. Because of the within-patch instability, the waves continually leave a wake of more-or-less empty patches behind them. The state of different patches thus becomes highly asynchronous, and the metapopulation as a whole now persists much more readily than its constituent parts could on their own.

The deterministic spatial patterns of dynamics that can be obtained by this process are striking and varied, and fall into three categories: 'spatial chaos', 'spiral waves' and 'crystal lattices' (Fig. 7.1). The spatial chaos is characterised by the populations fluctuating from patch to patch with no long-term spatial organisation; each randomly orientated wave front persists only briefly. The spiral waves differ in that the waves of local population densities tend to rotate in either direction around relatively immobile focal points. A time series taken from a particular local population would be characterised by fairly regular cycles produced by the regular 'passage' of the arm of the spiral through the habitat, except at the focus of the spiral where the population density would remain relatively constant. Local populations 'swept' by the trailing arms of the spiral fluctuate with increasing amplitude the further they are away from the focus. Strictly speaking, however, these patterns are also chaotic since the position and number of these focal points varies slowly in non-repeating patterns (Comins *et al.* 1992). The characteristic wavelength of these spirals can vary greatly depending on the parameters of the model. Finally, the so-

(a)

(b)

(c)

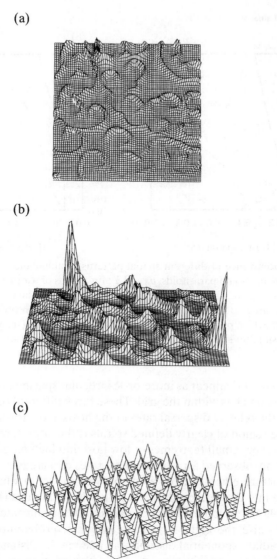

Fig. 7.1 Maps showing the spatial distribution of host and parasitoid population densities in a chosen generation from simulations of the metapopulation model in eqns (7.3) and (7.4) with Nicholson–Bailey local dynamics and host rate of increase, $\lambda = 2$, parasitoid searching efficiency, $a = 1$, and other parameters as specified below. In each case the arenas have reflective boundaries and the interactions are initiated with a single patch seeded with a small number of hosts and parasitoids and all other patches empty. (a) Spiral patterns obtained with host dispersal rate, $\mu_N = 0.4$ and parasitoid dispersal rate, $\mu_P = 0.04$; (b) chaotic pattern obtained with $\mu_N = 0.2$ and $\mu_P = 0.9$; (c) Lattice pattern obtained with $\mu_N = 0.01$ and $\mu_P = 0.89$. Arena sizes are 75×75 for (a) and (b) and 30×30 for (c).

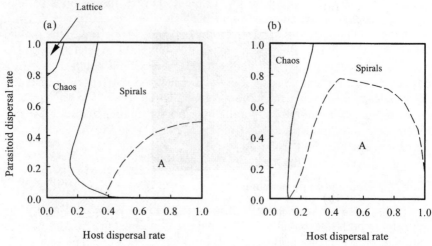

Fig. 7.2 The occurrence of different spatial patterns of abundance in relation to the dispersal rates of hosts and parasitoids, μ_N and μ_P, for arenas of width 30 and hosts rates of increase: (a) $\lambda = 2$ and (b) $\lambda = 10$. The boundaries are approximate and obtained by simulation. The area within the region of spirals labelled 'A' represents parameter combinations for which persistent spirals are difficult to establish (see p. 146), but once initiated using suitable starting conditions then readily persist. (After Comins *et al.* 1992.)

called 'crystal lattices' appear as more-or-less regular spacing of relatively high and low density patches within the grid. These three different patterns depend very much on the relative dispersal rates of the hosts and parasitoids, as shown in Fig. 7.2. The region of clearly defined spirals is the largest, while the crystal lattice region is very small (extremes of low host and high parasitoid dispersal rates), and is only observed for very low host rates of increase.

A similar picture emerges from a comparable model in which space is no longer divided into discrete patches, but rather is continuous (Kot *et al.* 1996; Wilson *et al.* 1997). Instead of hosts and parasitoids leaving patches at rates, μ_H and μ_P, to colonise the eight nearest neighbours, probability distributions (specifically, radial exponential distributions) define the distances likely to be moved by individuals. As with discrete space, the interactions readily persist, with different parameter combinations giving spiral and chaotic regions (but the crystal lattice area is absent, suggesting that it is a feature of discrete-space formulations).

How robust are all these results? Cellular automata with qualitative categories of patch occupancy give results that are qualitatively very similar to those of more detailed models (Hassell *et al.* 1991a). This suggests that the sorts of deterministic patterns in Fig. 7.1 are general to a whole class of predator–prey models and do not just depend on the detailed assumption of Nicholson–Bailey within-patch dynamics embedded within eqn (7.4). Theory is far ahead of experiment or observation in this instance. Direct observations of these kinds of dynamical spatial patterns in the field present enormous logistical problems. It may be possible, however, to determine properties of the population density distributions which are diagnostic of spirals or spatial chaos (for example, particular patterns of delayed covariance). This could then lead to indirect tests of the existence of these self-generated patterns in nature, that will also distinguish them from patterns of population abundance driven mainly by environmental randomness (paralleling work which tries to distinguish low-dimensional deterministic chaos from 'real noise' (Farmer and Sidorowich 1989; Sugihara and May 1990)).

7.3.3 Metapopulation persistence

One of the most striking features of metapopulations is the ease with which they tend to persist, even if the local populations are unstable, as long as there is sufficient asynchrony in the state of the different patches. The spatial patterns in Fig. 7.1 all promote such asynchrony, and are associated with characteristic types of total population fluctuation: large-amplitude irregular cycles from chaotic spatial patterns, smaller amplitude irregular cycles from spiral waves and stable populations from crystal lattices.

Although this persistence appears robust, there is always the possibility of extinction. Thus, extinction becomes increasingly likely as the size of the total habitat decreases (Fig. 7.3). In the limit of very small arena widths ($n < 3$) and all individuals dispersing from their respective patches ($\mu_N = \mu_P = 1$), local and metapopulation models become the same. For example, in the case of random parasitism within patches ($k \to \infty$), the Nicholson–Bailey model is recovered for the metapopulation as a whole and extinction therefore always occurs. As the arena size increases, the probability of extinction rapidly decreases. For a given size of arena, the host and parasitoid dispersal rates also influence the probability of extinction. Generally, as dispersal increases, adjacent patches become more synchronous and the host and parasitoid distributions become more autocorrelated in space. The characteristic spatial scale of the dynamics thus increases, and it becomes more difficult to fit the self-maintaining pattern into the available space. In much the same way, fragmentation of the whole environment runs the risk of disrupting the metapopulation dynamics, either by reducing the number of local populations below some level required for the combined metapopulation to persist, or by interfering with the dispersal required to link the locally unstable local populations (often leading to population outbreaks as the ability of the parasitoid to regulate the host is disrupted). The extent of this depends largely on the characteristic spatial scale of the dynamics. Thus parameter combinations producing large-scale spirals are especially vulnerable to shrinking arena sizes, while interactions producing chaotic spatial patterns are less vulnerable (Hassell *et al.* 1993).

But even with very large arenas, the metapopulation will not persist

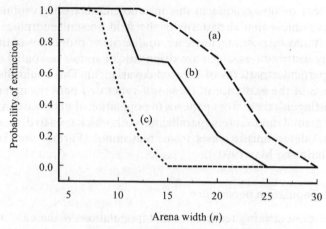

Fig. 7.3 Extinction probabilities for metapopulation simulations started in various sizes of arena of width n, for three different host dispersal rates: (a) $\mu_N = 0.1$, (b) $\mu_N = 0.4$ and (c) $\mu_N = 0.8$ (parasitoid dispersal rate, $\mu_P = 0.9$, and host rate of increase, $\lambda = 2$). Metapopulations in too small arenas never persist, while those in large arenas persist with a very high probability. (Comins *et al.* 1992.)

indefinitely. For convenience, 'persistence' in Fig. 7.2 has been arbitrarily defined as the population size of each species remaining above a certain extinction threshold for at least 2000 generations. This, however, hides the fact that there is always a small, constant probability of the entire metapopulation of hosts or parasitoids becoming extinct at each time step. Increasing the arena size (up to a threshold size) will decrease this probability of extinction, but can never eliminate it.

Finally, extinction may occur if the original 'invasion' of the area is on too small a scale or is otherwise unsuitable. For example, in region A in Fig. 7.2, persistence is impossible if the simulations are started from a single non-empty cell. However, once the populations are initiated by the simultaneous colonisation of many of the cells, persisting spirals are relatively easy to establish.

7.3.4 Variable environments and demographic stochasticity

The striking spatial patterns and persistence properties of metapopulations outlined above are observed from completely uniform and regular environments with no spatial or temporal variation in any of the parameters. This prompts the interesting question of how robust these properties are to various forms of random variation; in particular, to environmental and demographic stochasticity (May 1974b).

Environmental stochasticity. This is most easily represented by random variation in the demographic parameters from patch to patch (Reeve 1988;

Comins *et al.* 1992; Hassell *et al.* 1993; Ruxton and Rohani 1996). In this case there is little disruption of the spatial patterns observed from the corresponding deterministic cases, even with quite appreciable levels of random variation, primarily because the variation is on a much smaller spatial scale than the spatial patterns themselves (except for the crystal lattices that are easily disrupted). Only if the random noise is spatially correlated, with a scale of variation comparable to the characteristic scale of the spatial dynamics, does it markedly affect the outcome by disrupting the spatial patterns.

Demographic stochasticity. One feature of host–parasitoid metapopulations that have oscillatorily unstable local populations is the extremely low population size that can occur within patches, often represented by small fractions of an individual host or parasitoid. This raises the possibility that some of the properties described above are artefacts of allowing fractional individuals to occur at low population densities. For instance, the random, Nicholson–Bailey parasitism within patches is formulated as a mean field approximation of the random process of parasitoids finding hosts. While this is often a good approximation, it does not fare so well when there are very small numbers of individuals within patches (Pimm *et al.* 1988; Rand and Wilson 1991; Wilson *et al.* 1997). Secondly, and again when local population sizes are very small, some of the properties of the deterministic model may rely heavily on the equal dispersal to neighbouring patches of 'fractional individuals'. In short, demographic stochasticity affecting reproduction, parasitism and dispersal rates needs to be considered in the formulation of metapopulation models.

Let us consider an individual-based, host–parasitoid model in which encounters between hosts and parasitoids are modelled by a simple stochastic process (Wilson *et al.* 1997; Wilson and Hassell 1997). There are a number of benefits of doing this. First, because fractions of individuals cannot occur, dispersal at very low densities can become uneven. Second, the problem of the Nicholson–Bailey model being a poor approximation for random parasitism at very low densities is addressed. And third, a stochastic individual-based model allows random extinctions to be incorporated in a much more realistic way than simply by setting an arbitrary minimum size below which the local population is assumed to become extinct.

The full details of this stochastic individual-based model (SIB) are given in Wilson and Hassell (1997). It differs from the corresponding deterministic model (DET) in two important ways. First, random parasitism within each patch is modelled directly using a simple stochastic model (in the limiting case of no stochasticity, the Nicholson–Bailey model is exactly recovered). Second, each individual host and parasitoid disperses to a site randomly chosen from one of the neighbouring patches, with probability μ_H and μ_P, respectively. When the number of individuals within a patch is small, the dispersal process is not even, and there may, therefore, be no dispersal to some neighbouring patches. There are three broad differences between the properties of the SIB and DET models, as follows.

1. Since the effects of demographic stochasticity become increasingly marked as population size decreases, the parasitoid attack rate, a, has important affects in the SIB model that are not seen in the DET model, simply because mean host and parasitoid abundances per patch decrease at a rate proportional to $1/a$ (see p. 16). At small values of a both models behave similarly since the mean population sizes per patch tend to be large, and stochastic effects are averaged out. At larger values of a, however, the behaviour of the two models diverges. The properties of the DET model are unaltered, but in the SIB model the increased importance of stochastic effects means that extinctions of local host and parasitoid populations become more frequent. Population 'explosions' in individual patches thus occur with increasing frequency due to parasitoid extinctions thereby allowing hosts to increase unchecked. Finally, as a increases further, demographic stochasticity eventually causes the extinction of the parasitoid population in all the patches; something which is not seen in the DET model where mean population abundances can continue to decrease indefinitely at a rate proportional to $1/a$ (unless an arbitrary threshold for extinction is imposed).

2. With the exception of the 'crystal lattice' pattern (not seen in the stochastic model), the SIB model shows the same spatial patterns as found in the DET model, but with the chaos region greatly increased over that in Fig. 7.2 if the stochastic effects are large.

3. The effect of lattice size on the probability of extinction in the DET model is shown in Fig. 7.3 above. Broadly the same picture is obtained from the SIB model if the stochastic effects are small (i.e. small a), but very different if a is large. Now, low dispersal rates markedly decrease the probability of persistence in small lattices compared to the DET model in which eight neighbouring sites are always colonised, irrespective of the number of dispersers and even if this involves fractional numbers of individuals moving into a patch.

In short, small and intermediate levels of demographic stochasticity increase the region of spatial chaos and destabilise the crystal lattice structure found in the comparable deterministic models. They also cause a tendency for population explosions; the absence of 'fractional parasitoids' increases the frequency with which host patches remain undiscovered and, across the metapopulation as a whole, this effect becomes amplified to cause massive population fluctuations (akin to similar effects observed in some epidemiological models (Rand and Wilson 1991)). Finally, very large levels of demographic stochasticity cause the parasitoid population to go extinct and thus completely destabilise the dynamics.

This analysis suggests that the interplay between intrinsic dynamics and demographic stochasticity is likely to be an important feature of metapopulations in general whenever the probability of local extinction is quite high. It also affects the properties of more complex, multispecies metapopulations, as discussed in the next section.

7.4 Multispecies metapopulations

We now extend the basic two-species metapopulation to consider the dynamics of a range of three-species, host–parasitoid systems:

(1) two host species with a shared parasitoid species (H–P–H);
(2) a single host species attacked by two parasitoid species (P–H–P); and
(3) a host–parasitoid–hyperparasitoid interaction (H–P–Q);

(Hassell *et al.* 1994; Comins and Hassell 1996; Wilson and Hassell 1997; Bonsall and Hassell 2000). In particular, it will be interesting to determine if 'diffusive' dispersal in these more complicated metapopulations has the same profound effects of promoting persistence, and, if so, how this changes the conditions for the coexistence of competing species. The three types of interaction are modelled in much the same way as for the two-species, host–parasitoid systems. In each model generation there are again two phases: (1) the interactions of hosts and parasitoids within individual patches, described by the Nicholson–Bailey equations and (2) the dispersal of a fraction of the subsequently emerging adult hosts and parasitoids. Full details of the models are given in Comins and Hassell (1996). In the case of the P–H–P system, the two parasitoid species are assumed to search simultaneously for hosts within a patch, and superparasitism by either species is always assumed to be unsuccessful (Boerlijst *et al.* 1993).

All three models give similar results: for interactions that are always unstable in the limit of a single patch, a third species (be it another host, another parasitoid or a hyperparasitoid) can coexist for long periods within the spatial dynamics (spiral waves or chaos) generated by an existing two-species interaction. The conditions for doing so are restricted compared to the corresponding two-species system and depend upon some kind of fugitive coexistence (Hutchinson 1951; Horn and MacArthur 1971; Levins and Culver 1971; Hanski and Ranta 1983; Nee and May 1992; Hanski and Zhang 1993). For example, in the two parasitoid–one host system, coexistence occurs most easily when the two parasitoid species have very different dispersal rates, provided that the low dispersal rate is matched by high within-patch searching efficiency and vice versa.[1] Similarly, in the apparent competition case of two hosts and one shared parasitoid (Bonsall and Hassell, unpublished), coexistence occurs most readily when the two host species have very different dispersal rates and the relatively immobile species has either the higher rate of increase or is less susceptible to parasitism. Because of the lower parasitism and/or higher rate of increase, the superior apparent competitor occupies a greater fraction of the available patches—in which they are also locally more abundant. This there-

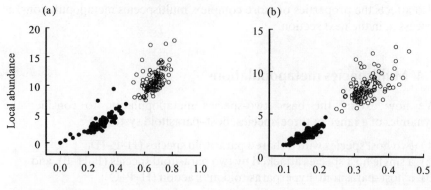

Fig. 7.4 The relationship between patch occupancy and local population abundance in three-species, host–parasitoid metapopulations. Results are for 200 simulations iterated for 1000 generations, in which all three species coexisted. (a) A two-host–one parasitoid model showing coexistence associated with the spatially more pervasive host (species X, solid circles) having a higher within-patch abundance than the spatially more restricted host (species Y, hollow circles). Parameters for host species X and Y and parasitoid P: host rates of increase, $\lambda_X = 2$, $\lambda_Y = 1.5$; parasitoid searching efficiency on both hosts, $a = 0.9$; dispersal rates, $\mu_X = \mu_P = 0.89$, $\mu_Y = 0.2$. (b) A two-parasitoid–one-host model showing coexistence now associated with the spatially more pervasive parasitoid (species P, solid circles) having the higher within-patch abundance compared to the more restricted parasitoid (species Q, hollow circles). Parameters for host species X and parasitoid species P and Q: $\lambda_X = 2$; $a_P = a_Q = 0.1$; $\mu_X = \mu_Q = 0.5$ and $\mu_P = 0.05$. (Bonsall and Hassell, unpublished.)

fore leads to the kind of strong positive correlation between patch occupancy and local abundance (Fig. 7.4) which has often been reported as a general observation independent of taxonomic group and habitat type (Hanski 1982; Brown 1984; Lawton 1993; Gaston *et al.* 1997). Finally, in the host–parasitoid–hyperparasitoid system, coexistence demands that the hyperparasitoid has a higher searching efficiency than the parasitoid (as concluded from non-spatial H–P–Q models by Beddington and Hammond (1977), Hassell (1979) and May and Hassell (1981)). As in the two-species models, introducing demographic stochasticity decreases the probability of persistence (Wilson *et al.* 1998).

Another interesting property of these models is that coexistence tends to be associated with some degree of self-organising niche separation between the competing species (Hassell *et al.* 1994; Comins and Hassell 1996; Bonsall and Hassell, unpublished). This is best seen when the spatial dynamics show clear spirals. For example, in the case of two competing parasitoids with very different dispersal rates, the relatively immobile species tends to be confined to the central foci of the spirals where it is the most abundant species, and the highly dispersive species occupies the remainder of the 'trailing arm' of the spirals, as shown in Fig. 7.5. Since the foci of spirals are relatively static in these models,

(a)   (b)

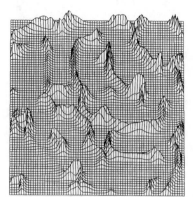
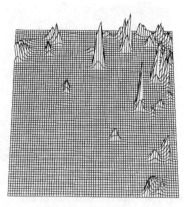

Fig. 7.5 Maps of the spatial distribution of densities of two parasitoid species from a chosen generation of a persistent parasitoid–host–parasitoid interaction with $\lambda = 2$, $a = 0.05$, host dispersal rate, $\mu_{\mathrm{H}} = 0.4$, parasitoid dispersal rates, $\mu_{\mathrm{P1}} = 0.05$ and $\mu_{\mathrm{P2}} = 0.5$, arena size of 75×75 and absorptive boundaries. (a) Distribution of the more dispersive parasitoid (P_1) and (b) of the more sedentary parasitoid (P_2). In the time-evolution of the system, the 'mountain ranges' of high population density in (a) are in continuous motion, while the peaks in (b) are at the foci of the spirals and are therefore relatively immobile. (Comins and Hassell 1996.)

the less mobile species appears to occur only in isolated, small 'islands' within the habitat, much as if these were pockets of favourable habitat. As the dispersal rates become less divergent between the species, the niche of the less dispersive species spreads further into the arm of the spirals.[2]

Another example, for a system with two host species and one parasitoid species, is shown in Fig. 7.6. In this case one host is 'superior' by virtue of a higher rate of increase and also has a higher dispersal rate than the 'inferior' host species. The superior host spreads throughout the habitat while the inferior species is confined to the margins and is unable to spread further. However, these intriguing properties appear to be quite sensitive to model conditions. For example, the segregation in Fig. 7.5 tends to disappear if there are threshold populations below which host and parasitoid cannot fall, and is not found at all in the SIB model above (Wilson and Hassell 1997); the 'focus-living' species gradually becomes eliminated from the foci, without being able to reinvade the vacant foci. Introduction of a small fraction of individuals capable of long-range dispersal, however, permits reinvasion of the vacant spiral foci and the phenomenon is once again observed (Comins and Hassell 1996).

The possibility that coexistence is promoted by trade-offs between competitive and dispersal abilities is well recognised. What is different here is that the spatial metapopulation structure, within which these trade-offs occur, may

(a)   (b)

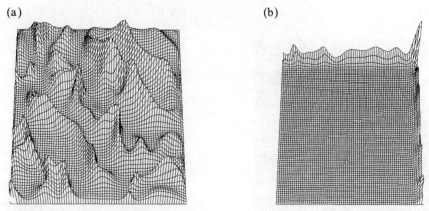

Fig. 7.6 Spatial maps as in Fig. 7.5, but now showing the segregated distribution of densities of two host species from a two-host–one-parasitoid interaction. (a) Distribution of host species X—the superior competitor—and (b) distribution of host species Y—the inferior competitor. Host X has a higher rate of increase than host Y ($\lambda_x = 1.5$, $\lambda_Y = 1.3$) and a higher dispersal rate ($\mu_x = 0.65, \mu_Y = 0.15$). The parasitoid species has a searching efficiency of $a = 0.1$ for both hosts and a dispersal rate of $\mu_x = 0.65$. (Bonsall and Hassell, unpublished.)

itself be self-generated by virtue of the dynamics of the interacting species, leading to competitors being separated in space in a self-organised way. An important challenge is to develop tests that will distinguish between species segregated in this way due to their internal dynamics from those whose distribution depends on a non-uniform environment.

7.5 Summary

This chapter reviews the dynamics of metapopulation models of discrete-generation, host–parasitoid interactions set in a grid of patches in which partially autonomous local populations reside. In a deterministic, two-species, host–parasitoid interaction with random parasitism within patches and global dispersal in each generation so that each patch is colonised with equal probability, local and metapopulations show the same unstable (Nicholson–Bailey) dynamics. Introducing restricted dispersal so that in each generation a fraction of hosts and parasitoids disperse to neighbouring patches, enables the metapopulation to persist with a high probability provided there are sufficient patches in the grid. Associated with this are characteristic patterns of spatial abundance, which may be chaotic, exhibit 'spiral waves' or take the form of a 'lattice' arrangement of high and low host density patches. The asynchronous dynamics of local populations that enables this persistence to occur can be promoted by sufficient levels of environmental and demographic stochasticity.

The basic host–parasitoid metapopulation models have been extended to include an additional host, parasitoid or hyperparasitoid species. The following conclusions emerge from these three-species systems:

1. Three-species persistence occurs in a restricted range of conditions compared to the two-species, host–parasitoid systems.
2. This persistence is enhanced by a 'balancing' of key parameters between the competing species (e.g. a high searching efficiency in one of the competitors associated with a lower dispersal rate).
3. Coexistence may be associated with some degree of persistent spatial segregation of the competing species, despite the completely uniform environment, although this is not seen in some models; for example, if there are threshold minima below which the populations cannot fall, or the interaction is formulated as a stochastic, individual-based model. It remains to be seen whether or not such 'self-organising' spatial segregation among competing species, imposed and maintained by population dynamics, is a real property of natural metapopulations.

Notes

1. There may, however, be constraints on coexistence that stem from the initial conditions. Boerlijst *et al.* (1993) have shown that where spiral waves occur, a relatively non-competitive parasitoid species may resist invasion by a superior species simply by being unable to cross the 'wasteland' between successive waves of the superior host's abundance.
2. This is an example, in yet another guise, of the dispersal/competition trade-off in metapopulations discussed, for example, by Nee *et al.* (1997) and Tilman *et al.* (1994, 1997).

8

Epilogue

The question of what regulates populations was first posed early in this century by entomologists, particularly Howard and Fiske (1911) and Thompson (1924) who were engaged in the design and execution of biological-control programmes, and by Nicholson (1933) who was interested in more general questions of population dynamics. Thompson and Nicholson's models were both framed in discrete generations and had random parasitoid encounters with hosts, but they made quite different assumptions about the efficiency of parasitoids in finding hosts. Thompson's parasitoids were always egg-limited and his models were inherently unstable. Nicholson's parasitoids had a potentially unlimited egg supply and were only limited by their searching ability, encapsulated in the searching efficiency parameter, the area of discovery. Ironically for someone who championed the cause of population regulation by density-dependent factors, his models were also unstable showing divergent oscillations in host and parasitoid population densities.

Nicholson's papers were immensely influential and established a baseline model for the interaction of coupled host and parasitoid populations. Much of the subsequent work on host–parasitoid dynamics has involved relaxing the assumptions of this basic model and, one by one, including more realistic components to determine how each affects dynamics. Several of these components have been reviewed in an earlier monograph (Hassell 1978) and are brought more up to date here in Chapters 2 and 3. Since then, there have been four important areas of development. First has been the recognition that not all hosts are equally susceptible to attack by parasitoids. All hosts tend to be distributed non-randomly across their environment, often in fairly discrete patches, and parasitoids search and locate hosts in this setting using an impressive variety of cues. Add to this the variations between individuals in their phenology, behaviour and host defences and it is inconceivable that the core assumption in the early models of equal susceptibility between individuals can apply in real systems. These developments are reviewed in Chapter 5. Second has been the inclusion of age-structure in host–parasitoid models. If both hosts and parasitoids have discrete, synchronised generations, as in the Nicholson–Bailey model, then there is little need for an explicit consideration of age-structure. However, if generations overlap then age-structure can become very important and produce interesting dynamical effects on the populations. These are reviewed in Chapter 4. Third has been the elaboration of the basic models to include additional species of hosts and parasitoids. This reveals interesting criteria for the coexistence of competing hosts or parasitoids (Chapter 6). Finally, the whole concept of local populations tending to mix thoroughly in each generation has been relaxed to allow the linking together of several local populations, with some restricted dispersal of individuals between the various units. This has drawn host–parasitoid studies into the currently burgeoning field of metapopulation dynamics and self-organising spatial structures (Chapter 7).

Although the theory of host–parasitoid interactions is relatively well advanced, the deficiencies, as in other areas of ecology, are in the integration of theory with data from real systems and in the subsequent hand-in-hand development of theory and observation. While many of the demographic components of host–parasitoid systems have been separately quantified by field measurement and laboratory experiment, there are still relatively few detailed long-term studies of insect populations and their natural enemies in which the essential life-table parameters have been measured *and* appropriate mechanistic models of the system developed. The great value of such long-term data lies in the relative ease with which one can then develop parameterised models of the system. Modern methods of analysis are making it much easier to parameterise relatively simple population models from detailed field studies, and this should open the way to a much closer integration of empirical and theoretical ecology. This approach differs from the more phenomenological time-series description of population data in that the initial development of the model is deductive—based on general ecological assumptions about the nature of the important demographic processes. The model parameters are then estimated from the data using non-linear estimation techniques. Subsequent development of the model depends upon its ability to explain and predict the behaviour of the real system. Recently, intermediate approaches, which incorporate reliable information on population processes into a time-series approach, have been developed (Ellner *et al.* 1998). However, in the development of parameterised population and community models it is the plant ecologists who are leading the way (Tilman 1981; Tilman and Sterner 1984; Pacala and Silander 1990; Watkinson 1990; Tilman and Wedin 1991; Pacala *et al.* 1993, 1996; Cain *et al.* 1995; Rees *et al.* 1996, 2000). There is a continuum from simple models of ideas (Pacala 1997), through deductive models of intermediate complexity incorporating the key elements of the population dynamics whose parameters are estimated from the available field data (Jones *et al.* 1993; Rees and Paynter 1997), ultimately through to systems where mechanistic models are designed around the planning of the field study, followed by data collection and parameter estimation carrying on simultaneously throughout the study (Pacala *et al.* 1993, 1996). Plants are easier to work with in this regard, but animal ecologists would do well to tread this path as much as possible.

This book has dealt very little with stochastic effects on population dynamics. Yet, environmental and demographic stochasticity can interact with non-linear regulatory processes in complicated ways. The result is that environmental noise can help or hinder the persistence of host–parasitoid associations, depending on the frequency spectrum of the noise and how it interacts with density-dependent mechanisms. Chesson (1986), Chesson and Case (1986) and others have made a start in codifying these effects in general, but much remains to be done. This is an area where laboratory microcosms should have an important role to play. Stochasticity can be manipulated precisely and the way that it affects the key demographic components can then be evaluated.

Another major development in which much crucial information is lacking is in the development of a population-based community ecology of parasitoids. Understanding parasitoid communities is likely to involve an understanding of the dynamics of oligophagous parasitoids: species whose dynamics can neither be decoupled from their hosts, nor understood by considering one host species in isolation. To do this will be difficult, but the simple and often strong trophic link between parasitoids and their hosts suggests that this research programme has a better chance of succeeding than many equivalent programmes revolving around other types of animal trophic relationship.

Finally, many people have speculated that heterogeneities are crucial to the persistence of host–parasitoid systems. At the local scale this involves heterogeneity in the host risk of parasitism, which can arise in many different ways. Of these, density-dependent aggregation by parasitoids in patches of high host density may not be as important as has been thought, but heterogeneity in host defences and aggregation irrespective of host density may be more important. Comprehending these processes in an explicitly spatial setting has been a preoccupation of theorists for a long time. But, despite much quantification of the heterogeneities themselves, there is still little *direct* empirical evidence, apart from in a few laboratory studies, pinpointing how these heterogeneities influence population dynamics in the field. Replicated, long-term, manipulation experiments in the field and laboratory will be invaluable in resolving this longstanding issue.

The integration of theory and observation is also much needed in studying heterogeneities at the metapopulation scale. Key questions revolve around the measurement (and types of analysis) needed to diagnose population density distributions indicative of spirals or spatial chaos, and methods of distinguishing self-generated spatial patterns from those arising due to environmental randomness. Direct observations at the metapopulation scale in the field present enormous logistical problems. But Hanski (1999) is showing the way from his field study of the Glanville fritillary metapopulation in SW Finland. This study is unique in the scale and detail of the observations and in the way that key parameters of colonisation and extinction are being measured and related to metapopulation theory.

References

Abrams, P. A. (1994). The fallacies of 'ratio-dependent' predation. *Ecology*, **75**, 1842–1850.

Adler, F. R. (1993). Migration alone can produce persistence of host–parasitoid models. *American Naturalist*, **141**, 642–650.

Akçakaya, H. R., Arditi, R., and Ginzburg, L. R. (1995). Ratio-dependent predation: an abstraction that works. *Ecology*, **76**, 995–1004.

Allen, J. C. (1975). Mathematical models of species interactions in time and space. *American Naturalist*, **109**, 319–342.

Anderson, J. F., Hoy, M. A., and Weseloh, R. M. (1977). Field cage assessment of the potential for establishment of *Rogas indiscretus* against the gypsy moth. *Environmental Entomology*, **6**, 365–380.

Anderson, R. M. (1978). The regulation of host population growth by parasitic species. *Parasitology*, **76**, 119–157.

Anderson, R. M. and May, R. M. (1978). Regulation and stability of host–parasite population interactions: I. Regulatory processes. *Journal of Animal Ecology*, **47**, 219–247.

Anderson, R. M. and May, R. M. (1993). *Infectious diseases of humans: dynamics and control*. Oxford University Press, Oxford.

Andrewartha, H. G. (1957). The use of conceptual models in population ecology. *Cold Spring Harbor Symposium on Quantitative Biology*, **22**, 219–236.

Andrewartha, H. G. and Birch, L. C. (1954). *The distribution and abundance of animals*. University of Chicago Press, Chicago, IL.

Anscombe, F. J. (1959). Sampling theory of the negative binomial and logarithmic series distributions. *Biometrika*, **37**, 358–382.

Arditi, R. (1983). A unified model of the functional response of predators and parasitoids. *Journal of Animal Ecology*, **52**, 293–303.

Arditi, R. and Akçakaya, H. R. (1990). Underestimation of the mutual interference of predators. *Oecologia*, **83**, 358–361.

Arditi, R. and Ginzburg, L. R. (1989). Coupling in predator–prey dynamics: ratio-dependence. *Journal of Theoretical Biology*, **139**, 311–326.

Arditi, R. and Glaizot, O. (1995). Assessing superparasitism with a model combining the functional response and the egg distribution of parasitoids. *Entomophaga*, **40**, 235–262.

Armstrong, R. A. and McGehee, R. (1980). Competitive exclusion. *American Naturalist*, **115**, 151–170.

Askew, R. R. (1961). On the biology of the inhabitants of oak galls of Cynipidae (Hymenoptera) in Britain. *Transactions of the Society for British Entomology*, **14**, 237–268.

Askew, R. R. (1971). Parasitic Insects. Heinemann, London.

Askew, R. R. and Shaw, M. R. (1979). Mortality factors affecting the leaf-mining stages of *Phyllonorycter* (Lepidoptera: Gracilariidae) on oak and birch. 2. Biology of parasite species. *Zoological Journal of the Linnean Society*, **67**, 51–64.

Askew, R. R. and Shaw, M. R. (1986). Parasitoid communities: their size, structure, and development. In *Insect parasitoids* (ed. J. K. Waage and D. Greathead), pp. 225–264. Academic Press, London.

Atkinson, W. D. and Shorrocks, B. (1984). Aggregation of larval Diptera over discrete and ephemeral breeding sites: the implications for coexistence. *American Naturalist*, **124**, 336–351.

Auslander, D. M., Oster, G. F., and Huffaker, C. B. (1974). Dynamics of interacting populations. *Journal of the Franklin Institute*, **297**, 345–376.

Bailey, V. A., Nicholson, A. J., and Williams, E. J. (1962). Interaction between hosts and parasites when some host individuals are more difficult to find than others. *Journal of Theoretical Biology*, **3**, 1–18.

Bakker, K. (1961). An analysis of factors which determine success in competition for food among larvae of *Drosophila melanogaster*. *Archives Neerlandaises de Zoologie*, **14**, 200–261.

Bakker, K., Bagchee, S. N., van Zwet, W. R., and Meelis, E. (1967). Host discrimination in *Pseudeucoila bochei* (Hymenoptera: Cynipidae). *Entomologia Experimentalis et Applicata*, **10**, 295–311.

Baltensweiler, W. (1984). The role of environment and reproduction in the population dynamics of the larch bud moth, *Zeiraphera diniana* Gn. (Lepidoptera, Tortricidae). *Advances in Invertebrate Reproduction*, **3**, 291–302.

Bauer, G. (1985). Population ecology of *Pardia tripunctatana* Schiff. and *Notocelia roborana* Den. and Schiff. (Lepidoptera, Tortricidae)—an example of 'equilibrium species'. *Oecologia*, **65**, 437–441.

Beddington, J. R. (1975). Mutual interference between parasites or predators and its effect on searching efficiency. *Journal of Animal Ecology*, **44**, 331–340.

Beddington, J. R. and Hammond, P. S. (1977). On the dynamics of host–parasite–hyperparasite interactions. *Journal of Animal Ecology*, **46**, 811–821.

Beddington, J. R., Free, C. A., and Lawton, J. H. (1975). Dynamic complexity in predator–prey models framed in difference equations. *Nature, London*, **255**, 58–60.

Beddington, J. R., Free, C. A., and Lawton, J. H. (1976). Concepts of stability and resilience in predator–prey models. *Journal of Animal Ecology*, **45**, 791–816.

Beddington, J. R., Free, C. A., and Lawton, J. H. (1978). Characteristics of successful natural enemies in models of biological control of insect pests. *Nature, London*, **273**, 513–519.

Begon, M., Sait, S. M., and Thompson, D. J. (1995). Persistence of a parasitoid–host system—refuges and generation cycles? *Proceedings of the Royal Society of London, Series B*, **260**, 131–137.

Begon, M., Sait, S. M., and Thompson, D. J. (1997). Two's company, three's a crowd: host–pathogen–parasitoid dynamics. In *Multitrophic interactions in terrestrial systems* (ed. A. C. Gange and V. K. Brown), pp. 307–332. Blackwell Science, Oxford.

Bellows, T. S. (1981). The descriptive properties of some models for density dependence. *Journal of Animal Ecology*, **50**, 139–156.

Bellows, T. S. (1982*a*). Analytical models for laboratory populations of *Callosobruchus chinensis* and *C. maculatus* (Coleoptera, Bruchidae). *Journal of Animal Ecology*, **51**, 263–287.

Bellows, T. S. (1982*b*). Simulation models for laboratory populations of *Callosobruchus chinensis* and *C. maculatus*. *Journal of Animal Ecology*, **51**, 597–623.

Bellows, T. S. and Hassell, M. P. (1984). Models for interspecific competition in laboratory populations of *Callosobruchus* spp. *Journal of Animal Ecology*, **53**, 831–848.

Bellows, T. S. and Hassell, M. P. (1988). The dynamics of age-structure host–parasitoid interactions. *Journal of Animal Ecology*, **57**, 259–268.

Benson, J. F. (1974). Population dynamics of *Bracon hebetor* Say (Hymenoptera: Braconidae) and *Ephestia cautella* (Walker). Lepidoptera: Phycitidae) in a laboratory ecosystem. *Journal of Animal Ecology*, **43**, 71–84.

Berdegue, M., Trumble, J. T., Hare, J. D., and Redak, R. A. (1996). Is it enemy-free space? The evidence for terrestrial insects and freshwater arthropods. *Ecological Entomology*, **21**, 203–217.

Bernstein, C., Kacelnik, A., and Krebs, J. R. (1991). Individual decisions and the distribution of predators in a patchy environment. II: The influence of travel costs and the structure of the environment. *Journal of Animal Ecology*, **60**, 205–226.

Berryman, A. A. (1992). The origins and evolution of predator–prey theory. *Ecology*, **73**, 1530–1535.

Berryman, A. A. (1996). What causes population cycles of forest Lepidoptera? *Trends in Ecology and Evolution*, **11**, 28–32.

Berryman, A. A. and Turchin, P. (1997). Detection of delayed density dependence: comment. *Ecology*, **78**, 318–320.

Bess, H. A., van den Bosch, R., and Haramoto, F. H. (1961). Fruit fly parasites and their activities in Hawaii. *Proceedings of the Hawaii Entomology Society*, **17**, 367–378.

Birch, L. C. (1971). The role of environmental heterogeneity and genetical heterogeneity in determining distribution and abundance. In *Dynamics of populations* (ed. P. J. Den Boer and G. R. Gradwell), pp. 109–128. Centre for Agricultural Publishing and Documentation, Wageningen.

Blank, T. H., Southwood, T. R. E., and Cross, D. J. (1967). The ecology of the partridge. 1. Outline population processes with particular reference to chick mortality and nest density. *Journal of Animal Ecology*, **37**, 549–556.

Bleasdale, J. K. A. and Nelder, J. A. (1960). Plant population and crop yield. *Nature, London*, **188**, 342.

Blythe, S. P., Nisbet, R. M., and Gurney, W. S. C. (1983). Formulating population models with differential ageing. In Population Biology (ed. H. I. Freedman and C. Strobeck), pp. 133–140. Springer-Verlag, Berlin.

Blythe, S. P., Nisbet, R. M., and Gurney, W. S. C. (1984). The dynamics of population models with distributed maturation periods. *Theoretical Population Biology*, **25**, 289–311.

Boerlijst, M. C., Lamers, M. E., and Hogeweg, P. (1993). Evolutionary consequences of spiral waves in a host–parasitoid system. *Proceedings of the Royal Society of London, B*, **253**, 15–18.

Bonsall, M. B. and Hassell, M. P. (1997). Apparent competition structures ecological assemblages. *Nature*, **388**, 371–373.

Bonsall, M. B. and Hassell, M. P. (1998). Population dynamics of apparent competition in a host–parasitoid assemblage. *Journal of Animal Ecology*, **67**, 918–929.

Bonsall, M. B. and Hassell, M. P. (1999). Parasitoid-mediated effects: apparent competition and the persistence of host–parasitoid assemblages. *Researches on Population Ecology*, **41**, 59–68.

Boswell, M. T. and Patil, G. P. (1970). Chance mechanisms generating the negative binomial distribution. In *Random counts in models and structures* (ed. G. P. Patil), pp. 1–22. Pennsylvania University Press, Pennsylvania, PA.

Bouletreau, M. (1986). The genetic and coevolutionary interactions between parasitoids and their hosts. In *Insect parasitoids* (ed. J. K. Waage and D. Greathead), pp. 169–200. Academic Press, London.

Briggs, C. J. (1993). Competition among parasitoid species on a stage-structured host and its effect on host suppression. *American Naturalist*, **141**, 372–397.

Briggs, C. J. and Godfray, H. C. J. (1996). The dynamics of insect–pathogen interactions in seasonal environments. *Theoretical Population Biology*, **50**, 149–177.

Briggs, C. J., Nisbet, R. M., and Murdoch, W. W. (1993). Coexistence of competing parasitoid species on a host with a variable life cycle. *Theoretical Population Biology*, **44**, 341–373.

Briggs, C. J., Murdoch, W. W., and Nisbet, R. M. (1999*a*). Recent developments in theory for biological control of insect pests by parasitoids. In *Theoretical approaches to biological control* (ed. B. A. Hawkins and H. V. Cornell), pp. 22–42. Cambridge University Press, Cambridge.

Briggs, C. J., Sait, S. M., Begon, M., Thompson, D. J., and Godfray, H. C. J. (2000). What causes generation cycles in cultures of stored product moths? *Journal of Animal Ecology*, **69** (In press.)

Brodmann, P. A., Wilcox, C. V., and Harrison, S. (1997). Mobile parasitoids may restrict the spatial spread of an insect outbreak. *Journal of Animal Ecology*, **66**, 65–72.

Brown, J. H. (1984). On the relationship between abundance and distribution of species. *American Naturalist*, **124**, 253–279.

Brown, M. W. and Cameron, E. A. (1979). Effects of disparlure and egg mass size on parasitism by the gypsy moth egg parasite, *Ooencyrtus kuwanai*. *Environmental Entomology*, **8**, 77–80.

Buckner, C. H. (1964). Metabolism, food consumption, and feeding behaviour in four species of shrews. *Canadian Journal of Zoology*, **42**, 259–279.

Buckner, C. H. (1969). The common shrew (*Sorex araneus*) as a predator of the winter moth (*Operophtera brumata*) near Oxford, England. *Canadian Entomologist*, **101**, 370–375.

Burnett, T. (1956). Effects of natural temperatures on oviposition of various numbers of an insect parasite (Hymenoptera, Chalcididae, Tenthredinidae). *Annals of the Entomological Society of America*, **49**, 55–59.

Burnett, T. (1958). A model of host–parasite interaction. *Proceedings of the 10th International Conference of Entomology*, **2**, 679–686.

Cain, M. L., Pacala, S. W., Silander, J. A., and Fortin, M. J. (1995). Neighborhood models of clonal growth in the white clover *Trifolium repens*. *American Naturalist*, **145**, 888–917.

Casas, J. and Hulliger, B. (1994). Statistical analysis of functional response experiments. *Biocontrol Science and Technology*, **4**, 133–145.

Casas, J., Gurney, W. S. C., Nisbet, R. M., and Roux, O. (1993). A probabalistic model for the functional response of a parasitoid at the behavioural time-scale. *Journal of Animal Ecology*, **62**, 194–204.

Caswell, H. (1978). Predator-mediated coexistence: a nonequilibrium model. *American Naturalist*, **112**, 127–154.

Charnov, E. L. (1976). Optimal foraging: the marginal value theorem. *Theoretical Population Biology*, **9**, 126–136.

Cheng, L. (1970). Timing of attack by *Lypha dubia* Fall. (Diptera: Tachinidae) on the winter moth *Operophtera brumata* L. (Lepidoptera: Geometridae) as a factor affecting parasite success. *Journal of Animal Ecology*, **39**, 313–320.

Chesson, P. L. (1978). Predator–prey theory and variability. *Annual Review of Ecology and Systematics*, **9**, 323–347.

Chesson, P. L. (1986). Environmental variation and the coexistence of species. In *Community ecology* (ed. J. Diamond and T. J. Case), pp. 240–256. Harper and Row, New York.

Chesson, P. L. and Case, T. J. (1986). Overview: nonequilibrium community theories: chance, variability, history, and coexistence. In *Community ecology* (ed. J. Diamond and T. J. Case), pp. 229–239. Harper and Row, New York.

Chesson, P. L. and Huntly, N. (1989). Short-term instabilities and long-term community dynamics. *Trends in Ecology and Evolution*, **4**, 293–298.

Chesson, P. L. and Murdoch, W. W. (1986). Aggregation of risk: relationships among host–parasitoid models. *American Naturalist*, **127**, 696–715.

Clark, L. R. (1963a). The influence of population density on the number of eggs laid by females of *Cardiaspina albitextura* (Psyllidae). *Australian Journal of Zoology*, **11**, 190–201.

Clark, L. R. (1963b). The influence of predation by *Syrphus* sp. on the numbers of *Cardiaspina albitextura* (Psyllidae). *Australian Journal of Zoology*, **11**, 470–487.

Clark, L. R. (1964a). The influence of parasite attack in relation to the abundance of *Cardiaspina albitextura* (Psyllidae). *Australian Journal of Zoology*, **12**, 150–173.

Clark, L. R. (1964b). The population dynamics of *Cardiaspina albitextura* (Psyllidae). *Australian Journal of Zoology*, **12**, 362–380.

Clausen, C. P. (1940). *Entomophagous insects*. Hafner, New York.

Clutton-Brock, T. H., Guiness, F., and Albon, S. (1982). Red deer: behaviour and ecology of two sexes. Edinburgh University Press, Edinburgh.

Cock, M. J. W. (1978). The assessment of preference. *Journal of Animal Ecology*, **47**, 805–816.

Comins, H. N. and Hassell, M. P. (1976). Predation in multi-prey communities. *Journal of Theoretical Biology*, **62**, 93–114.

Comins, H. N. and Hassell, M. P. (1979). The dynamics of optimally foraging predators and parasitoids. *Journal of Animal Ecology*, **48**, 335–351.

Comins, H. N. and Hassell, M. P. (1987). The dynamics of predation and competition in patchy environments. *Theoretical Population Biology*, **31**, 393–421.

Comins, H. N. and Hassell, M. P. (1996). Persistence of multispecies host–parasitoid interactions in spatially distributed models with local dispersal. *Journal of Theoretical Biology*, **183**, 19–28.

Comins, H. N., Hassell, M. P., and May, R. M. (1992). The spatial dynamics of host–parasitoid systems. *Journal of Animal Ecology*, **61**, 735–748.

Connell, J. H. and Sousa, W. P. (1983). On the evidence needed to judge ecological stability or persistence. *American Naturalist*, **121**, 789–824.

Cook, L. M. (1965). Oscillation in the simple logistic growth model. *Nature, London*, **207**, 316.

Cook, R. M. and Hubbard, S. F. (1977). Adapative searching strategies in insect parasites. *Journal of Animal Ecology*, **46**, 115–125.

Cornell, H. V. (1976). Search strategies and the adaptive significance of switching in some general predators. *American Naturalist*, **110**, 317–320.

Cramer, N. F. and May, R. M. (1972). Interspecific competition, predation and species diversity: a comment. *Journal of Theoretical Biology*, **34**, 289–290.

Crawley, M. J. (1975). The numerical responses of insect predators to changes in prey density. *Journal of Animal Ecology*, **44**, 877–892.

Crawley, M. J. (1992). Population dynamics of natural enemies and their prey. In *Natural enemies: the population biology of predators, parasites and diseases* (ed. M. J. Crawley), pp. 40–89. Blackwell Science, Oxford.

Crawley, M. J. and May, R. M. (1987). Population-dynamics and plant community structure—competition between annuals and perennials. *Journal of Theoretical Biology*, **125**, 475–489.

Crofton, H. D. (1971). A quantitative approach to parasitism. *Parasitology*, **62**, 179–194.

Cronin, J. T. and Strong, D. R. (1990). Density-independent parasitism among host patches by *Anagrus delicatus* (Hymenoptera: Mymaridae)—experimental manipulation of hosts. *Journal of Animal Ecology*, **59**, 1019–1026.

Cronin, J. T. and Strong, D. R. (1993). Superparasitism and mutual interference in the egg parasitoid *Anagrus delicatus* (Hymenoptera: Mymaridae). *Ecological Entomology*, **18**, 293–302.

Cronin, J. T. and Strong, D. R. (1999). Dispersal-dependent oviposition and the aggregation of parasitism. *American Naturalist*, **154**, 23–36.

Crowley, P. H. (1979). Predator-mediated coexistence: an equilibrium interpretation. *Journal of Theoretical Biology*, **80**, 129–144.

Crowley, P. H. (1981). Dispersal and the stability of predator–prey interactions. *American Naturalist*, **118**, 673–701.

DeBach, P. (1966). The competitive displacement and coexistence principles. *Annual Review of Entomology*, **11**, 183–212.

DeBach, P., Rosen, D., and Kennett, C. E. (1971). Biological control of coccids by introduced natural enemies. In *Biological control* (ed. C. B. Huffaker), pp. 165–194. Plenum Press, New York.

DeJong, G. (1979). The influence of the distribution of juveniles over patches of food on the dynamics of a population. *Netherlands Journal of Zoology*, **29**, 33–51.

DeJong, G. (1981). The influence of dispersal pattern on the evolution of fecundity. *Netherlands Journal of Zoology*, **32**, 1–30.

DeRoos, A. M., McCauley, E., and Wilson, W. G. (1998). Pattern formation and the spatial scale of interaction between predators and their prey. *Theoretical Population Biology*, **53**, 108–130.

Dempster, J. P. (1983). The natural control of populations of butterflies and moths. *Biological Reviews*, **58**, 461–481.

Den Boer, P. J. (1991). Seeing the wood for the trees: random walks or bounded fluctuations of populations? *Oecologia*, **79**, 143–149.

Dennis, B. and Taper, M. L. (1994). Density dependence in time series observations of natural populations—estimation and testing. *Ecological Monographs*, **64**, 205–224.

Dobson, A. P. and Hudson, P. J. (1992). Regulation and stability of a free-living host–parasite system: *Trichostrongylus tenuis* in red grouse. 2. Population models. *Journal of Animal Ecology*, **61**, 487–498.

Doutt, R. L. (1961). The dimensions of endemism. *Annals of the Entomological Society of America*, **54**, 46–53.

Driessen, G. and Hemerik, L. (1991). Aggregative responses of parasitoids and parasitism in populations of *Drosophila* breeding in fungi. *Oikos*, **61**, 96–107.

Driessen, G. and Hemerik, L. (1992). The time and egg budget of *Leptopilina clavipes*, a parasitoid of larval *Drosophila*. *Ecological Entomology*, **17**, 17–27.

Driessen, G. and Visser, M. E. (1997). Components of parasitoid interference. *Oikos*, **79**, 179–182.

Durrett, R. and Levin, S. A. (1994). Stochastic spatial models—a user's guide to ecological applications. *Philosophical Transactions of the Royal Society, London, Series B*, **343**, 329–350.

Ehler, L. E. (1987). Patch-exploitation efficiency in a torymid parasite of a gall midge. *Environmental Entomology*, **16**, 198–201.

Elliott, J. M. (1977). *Some methods for the statistical analysis of samples of benthic invertebrates*. Freshwater Biological Association, Ambleside.

Elliott, J. M. (1984). Numerical changes and population regulation in young migratory trout *Salmo trutta* in a Lake District stream. *Journal of Animal Ecology*, **53**, 327–350.

Ellner, S.P. and Turchin, P. (1995). Chaos in a noisy world: new methods and evidence from time- series analysis. *American Naturalist*, **145**, 343–375.

Ellner, S. P., Bailey, B. A., Bobashev, G. V., Gallant, A. R., Grenfell, B. T., and Nychka, D. W. (1998). Noise and nonlinearity in measles epidemics: combining mechanistic and statistical approaches to population modeling. *American Naturalist*, **151**, 425–440.

Elton, C. (1927). *Animal ecology*. Methuen, London.

Embree, D. G. (1965). The population dynamics of the winter moth in Nova Scotia 1954–82. *Memoirs of the Entomological Society of Canada*, **46**, 1–57.

Embree, D. G. (1966). The role of introduced parasites in the control of the winter moth in Nova Scotia. *Canadian Entomologist*, **98**, 1159–1168.

Embree, D. G. (1971). The biological control of the winter moth in Eastern Canada by introduced parasites. In *Biological control* (ed. C. B. Huffaker), pp. 217–225. Plenum Press, New York.

Evans, E. W. and England, S. (1996). Indirect interactions in biological control of insects: pests and natural enemies in alfalfa. *Ecological Applications*, **6**, 920–930.

Fan, Y. Q. and Petitt, F. L. (1994). Parameter estimation of the functional response. *Environmental Entomology*, **23**, 785–794.

Farmer, J. D. and Sidorowich, J. J. (1989). Exploiting chaos to predict the future and reduce noise. In *Evolution, learning and cognition* (ed. Y. C. Lee), pp. 277–304. World Scientific Press, New York.

Fisher, R. C. (1971). Aspects of the physiology of endoparasitic Hymenoptera. *Biological Reviews*, **46**, 243–278.

Force, D. C. (1970). Competition among four hymenopterous parasites of an endemic insect host. *Annals of the Entomological Society of America*, **63**, 1675–1688.

Force, D. C. (1974). Ecology of host–parasitoid communities. *Science*, **184**, 624–632.

Frank, J. H. (1967). The effect of pupal predators on a population of winter moth, *Operophtera brumata* (L.) (Hydriomenidae). *Journal of Animal Ecology*, **36**, 611–621.

Frank, S. A. (1985). Hierarchical selection theory and sex ratios. II. On applying the theory, and a test with fig wasps. *Evolution*, **39**, 949–964.

Free, C. A., Beddington, J. R., and Lawton, J. H. (1977). On the inadequacy of simple models of mutual interference for parasitism and predation. *Journal of Animal Ecology*, **46**, 543–554.

Freeland, W. J. (1983). Parasites and the coexistence of animal host species. *American Naturalist*, **121**, 223–236.

Fujii, K. (1977). Complexity–stability relationships of two-prey–one-predator species systems model: local and global stability. *Journal of Theoretical Biology*, **69**, 613–623.

Fujii, K. (1983). Resource dependent stability in an experimental laboratory resource–herbivore–carnivore system. *Researches on Population Ecology*, **3**, 15–26.

Gaston, K. J. and Lawton, J. H. (1987). A test of statistical techniques for detecting density dependence in sequential censuses of animal populations. *Oecologia*, **74**, 404–410.

Gaston, K. J. and Lawton, J. H. (1988). Patterns in body size, population dynamics, and regional distribution of bracken herbivores. *American Naturalist*, **132**, 662–680.

Gaston, K. J., Blackburn, T. M., and Lawton, J. H. (1997). Interspecific abundance range size relationships: an appraisal of mechanisms. *Journal of Animal Ecology*, **66**, 579–601.

Getz, W. M. (1984). Population dynamics: a unified approach. *Journal of Theoretical Biology*, **108**, 623–643.

Getz, W. M. and Mills, N. J. (1997). Host–parasitoid coexistence and egg-limited encounter rates. *American Naturalist*, **148**, 333–347.

Ginzburg, L. R. (1986). The theory of population dynamics: I. Back to first principles. *Journal of Theoretical Biology*, **122**, 385–399.

Ginzburg, L. R. and Akçakaya, H. R. (1992). Consequences of ratio-dependent predation for steady-state properties of ecosystems. *Ecology*, **73**, 1536–1543.

Glass, N. A. (1970). A comparison of two models of the functional response with emphasis on parameter estimation procedures. *Canadian Entomologist*, **102**, 1094–1101.

Gleeson, S. K. (1994). Density dependence is better than ratio dependence. *Ecology*, **75**, 1834–1835.

Godfray, H. C. J. (1994). *Parasitoids: behavioral and evolutionary ecology*. Princeton University Press, Princeton, NJ.

Godfray, H. C. J. and Chan, M. S. (1990). How insecticides trigger single-stage outbreaks in tropical pests. *Functional Ecology*, **4**, 329–337.

Godfray, H. C. J. and Hassell, M. P. (1987). Natural enemies can cause discrete generations in tropical insects. *Nature, London*, **327**, 144–147.

Godfray, H. C. J. and Hassell, M. P. (1988). The population biology of insect parasitoids. *Science Progress, Oxford*, **72**, 531–548.

Godfray, H. C. J. and Hassell, M. P. (1989). Discrete and continuous insect populations in tropical environments. *Journal of Animal Ecology*, **58**, 153–174.

Godfray, H. C. J. and Hassell, M. P. (1990). Encapsulation and host–parasitoid population dynamics. In *Parasitism: coexistence or conflict?* (ed. C. Toft), pp. 131–147. Oxford University Press, Oxford.

Godfray, H. C. J., and Müller, C. B. (1998). Host–parasitoid dynamics. In *Insects populations in theory and in practice* (ed. J. P. Dempster and I. F. G. McLean), pp. 135–165. Kluwer Academic, London.

Godfray, H. C. J. and Pacala, S. W. (1992). Aggregation and the population dynamics of parasitoids and predators. *American Naturalist*, **140**, 30–40.

Godfray, H. C. J. and Waage, J. K. (1991). Predictive modelling in biological control: the mango mealy bug (*Rastrococcus invadens*) and its parasitoids. *Journal of Applied Ecology*, **28**, 434–453.

Godfray, H. C. J., Hassell, M. P., and Holt, R. D. (1994). The dynamic consequences of the disruption of synchrony between hosts and parasitoids. *Journal of Animal Ecology*, **63**, 1–10.

Gordon, D. M. (1987). Population dynamics of a host–parasitoid system with particular reference to age-structured effects. PhD thesis. McGill University, Montreal, Canada.

Gordon, D. M., Gurney, W. S. C., Nisbet, R. M., and Stewart, R. K. (1988). A model of *Cadra cautella* larval growth and development. *Journal of Animal Ecology*, **57**, 645–658.

Gordon, D. M., Nisbet, R. M., De Roos, A., Gurney, W. S. C., and Stewart, R. K. (1991). Discrete generations in host–parasitoid models with contrasting life cycles. *Journal of Animal Ecology*, **60**, 295–308.

Gotelli, N. J. (1995). *A primer of ecology*. Sinauer Associates, Sunderland, MA.

Gould, J. R., Elkington, J. S., and Wallner, W. E. (1990). Density-dependent suppression of experimentally created gypsy moth, *Lymantria dispar* (Lepidoptera: Lymantriidae), populations by natural enemies. *Journal of Animal Ecology*, **59**, 213–233.

Graham, A. R. (1958). Recoveries of introduced species of parasites of the winter moth, *Operophtera brumata* (L.) (Lepidoptera: Geometridae), in Nova Scotia. *Canadian Entomologist*, **90**, 595–596.

Greathead, D. J. and Greathead, A. H. (1992). Biological control of insect pests by insect parasitoids and predators: the BIOCAT database. *Biocontrol News and Information*, **13**, 61–68.

Green, R. F. (1986). Does aggregation prevent competitive exclusion? A response to Atkinson and Shorrocks. *American Naturalist*, **128**, 301–304.

Griffiths, K. J. (1969*a*). Development and diapause in *Pleolophus basizonus* (Hymenoptera: Ichneumonidae). *Canadian Entomologist*, **101**, 907–914.

Griffiths, K. J. (1969*b*). The importance of coincidence in the functional and numerical responses of two parasites of the European pine sawfly, *Neodiprion sertifer*. *Canadian Entomologist*, **101**, 673–713.

Griffiths, K. J. and Holling, C. S. (1969). A competition submodel for parasites and predators. *Canadian Entomologist*, **101**, 785–818.

Grover, J. (1990). Resource competition in a variable environment: phytoplankton growing according to Monod's model. *American Naturalist*, **136**, 771–789.

Grover, J. (1991). Resource competition in a variable environment: phytoplankton growing according to the variable-stores model. *American Naturalist*, **138**, 811–835.

Gurney, W. S. C. and Nisbet, R. M. (1978). Single species population fluctuations in patchy environments. *American Naturalist*, **112**, 1075–1090.

Gurney, W. S. C. and Nisbet, R. M. (1985). Fluctuation periodicity, generation separation, and the expression of larval competition. *Theoretical Population Biology*, **28**, 150–180.

Gurney, W. S. C., Nisbet, R. M., and Lawton, J. H. (1983). The systematic formulation of tractable single-species population models. *Journal of Animal Ecology*, **52**, 479–496.

Gyllenberg, M. and Hanski, I. (1997). Habitat deterioration, habitat destruction, and metapopulation persistence in a heterogenous landscape. *Theoretical Population Biology*, **52**, 198–215.

Hails, R. (1989). Host size and sex allocation of parasitoids in a gall forming community. *Oecologia*, **81**, 28–32.

Hails, R. and Crawley, M. J. (1992). Spatial density dependence in populations of a cynipid gall-former *Andricus quercuscalicis*. *Journal of Animal Ecology*, **61**, 567–583.

Hairston, N. G., Smith, F. E., and Slobodkin, L. E. (1960). Community structure, population control, and competition. *American Naturalist*, **94**, 421–425.

Hamilton, W. D. (1967). Extraordinary sex ratios. *Science*, **156**, 477–488.

Hanski, I. (1982). Dynamics of regional distribution: the core and satellite species hypothesis. *Oikos*, **38**, 210–221.

Hanski, I. (1987). Pine sawfly population dynamics: patterns, processes, problems. *Oikos*, **50**, 327–335.

Hanski, I. (1990). Density dependence, regulation and variability in animal populations. *Philosophical Transactions of the Royal Society, London, Series B*, **330**, 141–150.

Hanski, I. (1991). Single species metapopulation dynamics: concepts, models and observations. *Biological Journal of the Linnean Society*, **42**, 17–38.

Hanski, I. (1994). A practical model of metapopulation dynamics. *Journal of Animal Ecology*, **63**, 151–162.

Hanski, I. (1996). Habitat destruction and metapopulation dynamics. In *Enhancing the ecological basis of conservation: heterogeneity, ecosystem function and biodiversity* (ed. S. T. A. Pickett, R. A. Ostfeld, M. Shachak, and G. E. Likens), pp. 13–43. Chapman and Hall, New York.

Hanski, I. (1997a). Metapopulation dynamics: from concepts and observations to predictive models. In *Metapopulation biology: ecology, genetics, and evolution* (ed. I. Hanski and M. E. Gilpin), pp. 69–72. Academic Press, San Diego, CA.

Hanski, I. (1997b). Predictive and practical metapopulation models: the incidence function approach. In *Spatial ecology: the role of space in population dynamics and interspecific interactions* (ed. D. Tilman and P. Karieva), pp. 21–45. Princeton University Press, Princeton, NJ.

Hanski, I. (1999). *Metapopulation ecology*. Oxford University Press, Oxford.

Hanski, I. and Kuussaari, M. (1995). Butterfly metapopulation dynamics. In *Population dynamics* (ed. N. Cappuccino and P. W. Price), pp. 149–171. Academic Press, New York.

Hanski, I. and Parviainen, P. (1985). Cocoon predation by small mammals, and pine sawfly population-dynamics. *Oikos*, **45**, 125–136.

Hanski, I. and Ranta, E. (1983). Coexistence in a patchy environment: three species of *Daphnia* in rock pools. *Journal of Animal Ecology*, **52**, 263–279.

Hanski, I. and Simberloff, D. (1997). The metapopulation approach, its history, conceptual domain, and application to conservation. In *Metapopulation biology. Ecology, genetics, and evolution* (ed. I. Hanski and M. E. Gilpin), pp. 5–26. Academic Press, San Diego, CA.

Hanski, I. and Woiwod, I. P. (1991). Delayed density dependence. *Natural Resource Modeling*, **350**, 28.

Hanski, I. and Woiwod, I. P. (1993). Spatial synchrony in the dynamics of moth and aphid populations. *Journal of Animal Ecology*, **62**, 656–668.

Hanski, I. and Zhang, D. Y. (1993). Migration, metapopulation dynamics and fugitive co-existence. *Journal of Theoretical Biology*, **163**, 491–504.

Hanski, I., Woiwod, I. P., and Perry, J. N. (1993). Density dependence, population persistence, and largely futile arguments. *Oecologia*, **95**, 595–598.

Hanski, I., Moilanen, A., Pakkala, T., and Kuussaari, M. (1995a). Metapopulation persistence of an endangered butterfly: a test of the quantitative incidence function model. *Conservation Biology*, **10**, 578–590.

Hanski, I., Pakkala, T., Kuussaari, M., and Lei, G. C. (1995b). Metapopulation persistence of an endangered butterfly in a fragmented landscape. *Oikos*, **72**, 21–28.

Hanski, I., Moilanen, A., Pakkala, T., and Kuussaari, M. (1996). The quantitative incidence function model and persistence of an endangered butterfly metapopulation. *Conservation Biology*, **10**, 578–590.

Harcourt, D. G. (1971). Population dynamics of *Leptinotarsa decemlineata* (Say) in eastern Ontario. III. Major population processes. *Canadian Entomologist*, **103**, 1049–1061.

Hardy, I. C. W., Griffiths, N. T., and Godfray, H. C. J. (1992). Clutch size in a parasitoid wasp: a manipulation experiment. *Journal of Animal Ecology*, **61**, 121–129.

Harper, J. L. (1977). *The population biology of plants*. Academic Press, London.

Harrison, S. (1991). Local extinction in a metapopulation context: an empirical evaluation. In *Metapopulation dynamics: empirical and theoretical investigations* (ed. M. E. Gilpin and I. Hanski), pp. 73–88. Academic Press, London.

Harrison, S. and Cappuccino, N. (1995). Using density manipulation experiments to study population regulation. In *Population dynamics* (ed. N. Cappuccino and P. W. Price), pp. 131–147. Academic Press, New York.

Harrison, S. and Taylor, A. D. (1997). Empirical evidence for metapopulation dynamics. In *Metapopulation biology. Ecology, genetics, and evolution* (ed. I. Hanski and M. E. Gilpin), pp. 27–42. Academic Press, San Diego, CA.

Harrison, S., Murphy, D. D., and Ehrlich, P. R. (1988). Distribution of the Bay Checkerspot butterfly, *Euphydryas editha bayensis*: evidence for a metapopulation model. *American Naturalist*, **132**, 360–382.

Hassell, D. C., Allwright, D. J., and Fowler, A. C. (1999). A mathematical analysis of Jones's site model for spruce budworm infestations. *Journal of Mathematical Biology*, **38**, 377–421.

Hassell, M. P. (1968). The behavioural response of a tachinid fly *Cyzenis albicans* (Fall.) to its host, the winter moth, *Operophtera brumata* (L.). *Journal of Animal Ecology*, **37**, 627–639.

Hassell, M. P. (1969a). A population model for the interaction between *Cyzenis albicans* (Fall.) (Tachinidae) and *Operophtera brumata* (L.). Geometridae) at Wytham, Berkshire. *Journal of Animal Ecology*, **38**, 567–576.

Hassell, M. P. (1969b). A study of the mortality factors acting upon *Cyzenis albicans*

(Fall.), a tachinid parasite of the winter moth (*Operophtera brumata* (L.)). *Journal of Animal Ecology*, **38**, 329–339.

Hassell, M. P. (1975). Density dependence in single-species populations. *Journal of Animal Ecology*, **44**, 283–295.

Hassell, M. P. (1978). *The dynamics of arthropod predator–prey systems*. Princeton University Press, Princeton, NJ.

Hassell, M. P. (1979). The dynamics of predator–prey interactions: polyphagous predators, competing predators and hyperparasitoids. In *Population dynamics* (ed. R. M. Anderson, B. D. Turner, and L. R. Taylor), pp. 283–306. Blackwell Science, Oxford.

Hassell, M. P. (1980*a*). Foraging strategies, population models and biological control: a case study. *Journal of Animal Ecology*, **49**, 603–628.

Hassell, M. P. (1980*b*). Some consequences of habitat heterogeneity for population dynamics. *Oikos*, **35**, 150–160.

Hassell, M. P. (1982). Patterns of parasitism by insect parasitoids in patchy environments. *Ecological Entomology*, **7**, 365–377.

Hassell, M. P. (1984a). Parasitism in patchy environments: inverse density dependence can be stabilizing. *IMA Journal of Mathematics Applied in Medicine and Biology*, **1**, 123–133.

Hassell, M. P. (1984b). Insecticides in host-parasitoid interactions. *Theoretical Population Biology*, **26**, 378–386.

Hassell, M. P. (1985). Insect natural enemies as regulating factors. *Journal of Animal Ecology*, **54**, 323–334.

Hassell, M. P. (1986). Detecting density dependence. *Trends in Ecology and Evolution*, **1**, 90–93.

Hassell, M. P. (1987). Detecting regulation in patchily distributed animal populations. *Journal of Animal Ecology*, **56**, 705–713.

Hassell, M. P. and Anderson, R. M. (1984). Host susceptibility as a component in host–parasitoid systems. *Journal of Animal Ecology*, **53**, 611–621.

Hassell, M. P. and Anderson, R. M. (1989). Predator–prey and host–pathogen interactions. In *Ecological concepts: the contribution of ecology to an understanding of the natural world* (ed. J. M. Cherrett), pp. 147–196. Blackwell Science, Oxford.

Hassell, M. P. and Comins, H. N. (1975). Discrete time models for two-species competition. *Theoretical Population Biology*, **9**, 202–221.

Hassell, M. P. and Comins, H. N. (1978). Sigmoid functional responses and population stability. *Theoretical Population Biology*, **14**, 62–66.

Hassell, M. P. and Huffaker, C. B. (1969). Regulatory processes and population cyclicity in laboratory populations of *Anagasta kühniella* (Zeller). Lepidoptera: Phycitidae). III. The development of population models. *Researches on Population Ecology*, **11**, 186–210.

Hassell, M. P. and May, R. M. (1973). Stability in insect host–parasite models. *Journal of Animal Ecology*, **42**, 693–726.

Hassell, M. P. and May, R. M. (1974). Aggregation of predators and insect parasites and its effect on stability. *Journal of Animal Ecology*, **43**, 567–594.

Hassell, M. P. and May, R. M. (1985). From individual behaviour to population dynamics. In *Behavioural ecology* (ed. R. M. Sibly and R. Smith), pp. 3–32. Blackwell Science, Oxford.

Hassell, M. P. and May, R. M. (1986). Generalist and specialist natural enemies in insect predator–prey interactions. *Journal of Animal Ecology*, **55**, 923–940.

Hassell, M. P. and May, R. M. (1988). Spatial heterogeneity and the dynamics of parasitoid–host systems. *Annals Zoologici Fennici*, **25**, 55–61.

Hassell, M. P. and Pacala, S. W. (1990). Heterogeneity and the dynamics of host–parasitoid interactions. *Philosophical Transactions of the Royal Society, London, Series B*, **330**, 203–220.

Hassell, M. P. and Rogers, D. J. (1972). Insect parasite responses in the development of population models. *Journal of Animal Ecology*, **41**, 661–676.

Hassell, M. P. and Varley, G. C. (1969). New inductive population model for insect parasites and its bearing on biological control. *Nature, London*, **223**, 1133–1137.

Hassell, M. P., Lawton, J. H., and Beddington, J. R. (1976a). The components of arthropod predation. I. The prey death rate. *Journal of Animal Ecology*, **45**, 135–164.

Hassell, M. P., Lawton, J. H., and May, R. M. (1976b). Patterns of dynamical behaviour in single species populations. *Journal of Animal Ecology*, **45**, 471–486.

Hassell, M. P., Lawton, J. H., and Beddington, J. R. (1977). Sigmoid functional responses by invertebrate predators and parasitoids. *Journal of Animal Ecology*, **46**, 249–262.

Hassell, M. P., Waage, J. K., and May, R. M. (1983). Variable parasitoid sex ratios and their effect on host parasitoid dynamics. *Journal of Animal Ecology*, **52**, 889–904.

Hassell, M. P., Southwood, T. R. E., and Reader, P. M. (1987). The dynamics of the viburnum whitefly (*Aleurotrachelus jelinekii*): a case study of population regulation. *Journal of Animal Ecology*, **56**, 283–300.

Hassell, M. P., Taylor, V. A., and Reader, P. M. (1989). The dynamics of laboratory populations of *Callosobruchus chinensis* and *C. maculatus* in patchy environments. *Researches on Population Ecology*, **31**, 35–52.

Hassell, M. P., Comins, H. N., and May, R. M. (1991a). Spatial structure and chaos in insect population dynamics. *Nature, London*, **353**, 255–258.

Hassell, M. P., Pacala, S. W., May, R. M., and Chesson, P. L. (1991b). The persistence of host–parasitoid associations in patchy environments. I. A general criterion. *American Naturalist*, **138**, 568–583.

Hassell, M. P., Godfray, H. C. J., and Comins, H. N. (1993). Effects of global change on the dynamics of insect host–parasitoid interactions. In *Biotic interactions and global change* (ed. P. M. Karieva, J. G. Kingsolver, and R. B. Huey), pp. 402–423. Sinauer Associates, Sunderland, MA.

Hassell, M. P., Comins, H. N., and May, R. M. (1994). Species coexistence and self-organizing spatial dynamics. *Nature, London*, **370**, 290–292.

Hassell, M. P., Miramontes, O., Rohani, P., and May, R. M. (1995). Appropriate formulations for dispersal in spatially structured models: comments on Bascompte and Solé. *Journal of Animal Ecology*, **64**, 662–664.

Hassell, M. P., Crawley, M. J., Godfray, H. C. J., and Lawton, J. H. (1998). Top-down versus bottom-up and the Ruritanian bean bug. *Proceedings of the National Academy of Sciences USA*, **95**, 10661–10664.

Hastings, A. (1978). Spatial heterogeneity and the stability of predator–prey systems; predator-mediated coexistence. *Theoretical Population Biology*, **14**, 380–395.

Hastings, A. (1983). Age-dependent predation is not a simple process. 1. Continuous-time models. *Theoretical Population Biology*, **23**, 347–362.

Hastings, A. (1984). Age-dependent predation is not a simple process. II. Wolves, ungulates, and a discrete time model for predation on juveniles with a stabilizing tail. *Theoretical Population Biology*, **26**, 271–282.

Hastings, A. (1990). Spatial heterogeneity and ecological models. *Ecology*, **71**, 426–428.

Hastings, A. and Godfray, H. C. J. (1999). Learning, host fidelity, and the stability of host–parasitoid communities. *The American Naturalist*, **153**, 295–301.

Hastings, A., Harrison, S., and McCann, K. (1997). Unexpected spatial patterns in an insect outbreak match a predator diffusion model. *Proceedings of the Royal Society of London, Series B*, **264**, 1837–1840.

Hawkins, B. A. (1988). Species-diversity in the 3rd and 4th trophic levels—patterns and mechanisms. *Journal of Animal Ecology*, **57**, 137–162.

Hawkins, B. A. (1992). Parasitoid–host food webs and donor control. *Oikos*, **65**, 159–162.

Hawkins, B. A. and Goeden, R. D. (1984). Organization of a parasitoid community associated with a complex of galls on *Atriplex* spp. in Southern-California. *Ecological Entomology*, **9**, 271–292.

Hawkins, B. A. and Lawton, J. H. (1987). Species richness for parasitoids of British phytophagous insects. *Nature*, **326**, 788–790.

Hawkins, B. A., Thomas, M. B., and Hochberg, M. E. (1993). Refuge theory and biological control. *Science*, **262**, 1429–1432.

Herre, E. A. (1985). Sex ratio adjustment in fig wasps. *Science*, **228**, 896–898.

Herre, E. A. (1987). Optimality, plasticity and selective regime in fig wasp sex ratios. *Nature, London*, **329**, 627–629.

Hilborn, R. (1975). The effect of spatial heterogeneity on the persistence of predator–prey interactions. *Theoretical Population Biology*, **8**, 346–355.

Hochberg, M. E. and Hawkins, B. A. (1992). Refuges as a predictor of parasitoid diversity. *Science*, **255**, 973–976.

Hochberg, M. E. and Hawkins, B. A. (1993). Predicting parasitoid species richness. *American Naturalist*, **142**, 671–693.

Hochberg, M. E. and Holt, R. D. (1995). Refuge evolution and the population-dynamics of coupled host–parasitoid associations. *Evolutionary Ecology*, **9**, 633–661.

Hochberg, M. E. and Lawton, J. H. (1990). Spatial heterogeneities in parasitism and population dynamics. *Oikos*, **59**, 9–14.

Hochberg, M. E. and Waage, J. K. (1991). A model for the biological control of *Oryctes rhinoceros* (Coleoptera, Scarabaeidae) by means of pathogens. *Journal of Applied Ecology*, **28**, 514–531.

Hochberg, M. E., Hassell, M. P., and May, R. M. (1990). The dynamics of host–parasitoid–pathogen interactions. *American Naturalist*, **135**, 74–94.

Hogarth, W. L. and Diamond, P. (1984). Interspecific competition in larvae between entomophagous parasitoids. *American Naturalist*, **124**, 552–560.

Holling, C. S. (1959a). The components of predation as revealed by a study of small mammal predation of the European pine sawfly. *Canadian Entomologist*, **91**, 293–320.

Holling, C. S. (1959b). Some characteristics of simple types of predation and parasitism. *Canadian Entomologist*, **91**, 385–398.

Holling, C. S. (1961). Principles of insect predation. *Annual Review of Entomology*, **6**, 163–182.

Holling, C. S. (1973). Resilience and stability of ecological systems. *Annual Review of Ecology and Systematics*, **4**, 1–23.

Holt, R. D. (1977). Predation, apparent competition and the structure of prey communities. *Theoretical Population Biology*, **12**, 197–229.

Holt, R. D. (1984). Spatial heterogeneity, indirect interactions, and the coexistence of species. *American Naturalist*, **124**, 377–406.

Holt, R. D. (1997*a*). Community modules. In *Multitrophic interactions in terrestrial systems* (ed. A. C. Gange and V. K. Brown), pp. 333–350. Blackwell Science, Oxford.

Holt, R. D. (1997*b*). From metapopulation dynamics to community structure: some consequences of spatial heterogeneity. In *Metapopulation biology. Ecology, genetics, and evolution* (ed. I. Hanski and M. E. Gilpin), pp. 149–164. Academic Press, San Diego, CA.

Holt, R. D. and Hassell, M. P. (1993). Environmental heterogeneity and the stability of host–parasitoid interactions. *Journal of Animal Ecology*, **62**, 89–100.

Holt, R. D. and Kotler, B. P. (1987). Short-term apparent competition. *American Naturalist*, **130**, 412–430.

Holt, R. D. and Lawton, J. H. (1993). Apparent competition and enemy-free space in insect host–parasitoid communities. *American Naturalist*, **142**, 623–645.

Holt, R. D. and Lawton, J. H. (1994). The ecological consequences of shared natural enemies. *Annual Review of Ecology and Systematics*, **25**, 495–520.

Holt, R. D., Grover, J., and Tilman, D. (1994). Simple rules for interspecific dominance in systems with exploitative and apparent competition. *American Naturalist*, **144**, 741–771.

Holt, R. D., Lawton, J. H., Gaston, K. J., and Blackburn, T. M. (1997). On the relationship between range size and local abundance: back to basics. *Oikos*, **78**, 183–190.

Holyoak, M. (1993). The frequency of detection of density dependence in insect orders. *Ecological Entomology*, **18**, 339–347.

Holyoak, M. (1994*a*). Appropriate time scales for identifying lags in density-dependent processes. *Journal of Animal Ecology*, **63**, 479–483.

Holyoak, M. (1994*b*). Identifying delayed density dependence in time-series data. *Oikos*, **79**, 296–304.

Holyoak, M. and Lawler, S. P. (1996). Persistence of an extinction-prone predator–prey interaction through metapopulation dynamics. *Ecology*, **77**, 1867–1879.

Hopkins, M. J. G. (1984). The parasitoid complex associated with stem boring *Apion* (Col.: Curculionidae) feeding on *Rumex* species (Polygonaceae). *Entomologist's Monthly Magazine*, **120**, 187–192.

Hopper, J. L. (1990). Opportunities and handicaps of Antipodean scientists: A. J. Nicholson and V. A. Bailey on the balance of animal populations. *Historical Records of Australian Science*, **7**, 179–188.

Horn, H. S. and MacArthur, R. H. (1971). Competition among fugitive species in a harlequin environment. *Ecology*, **53**, 749–752.

Howard, L. O. and Fiske, W. F. (1911). The importation into the United States of the parasites of the gypsy-moth and the brown-tail moth. *Bulletin of the Bureau of Entomology, US Department of Agriculture*, **91**, 1–312.

Hubbard, S. F. and Cook, R. M. (1978). Optimal foraging by parasitoid wasps. *Journal of Animal Ecology*, **47**, 593–604.

Hudson, P. J., Newborn, D., and Dobson, A. P. (1992). Regulation and stability of a free-living host–parasite system—*Trichostrongylus tenuis* in red grouse. 1. Monitoring and parasite reduction experiments. *Journal of Animal Ecology*, **61**, 477–486.

Huffaker, C. B. (1958). Experimental studies on predation: dispersion factors and predator–prey oscillations. *Hilgardia*, **27**, 343–383.

Huffaker, C. B., Shea, K. P., and Herman, S. G. (1963). Experimental studies on predation. Complex dispersion and levels of food in an acarine predator–prey interaction. *Hilgardia*, **34**, 305–329.

Huffaker, C. B., Messenger, P. S., and De Bach, P. (1971). The natural enemy component in natural control and the theory of biological control. In *Biological control* (ed. C. B. Huffaker), pp. 16–67. Plenum, New York.

Hunter, M. D. and Price, P. W. (1992). Playing chutes and ladders: heterogeneity and the relative roles of bottom-up and top-down forces in natural communities. *Ecology*, **73**, 724–732.

Hutchinson, G. E. (1951). Copepodology for the ornithologist. *Ecology*, **32**, 571–577.

Ibarra, E. L., Wallwork, J. A., and Rodriguez, J. G. (1965). Ecological studies of mites found in sheep and cattle pastures. I. Distribution patterns of oribatid mites. *Annals of the Entomological Society of America*, **58**, 153–159.

Ito, Y. (1980). *Comparative ecology* (trans. J. Kikkawa). Cambridge University Press, Cambridge.

Ives, A. R. (1992a). Continuous-time models of host–parasitoid interactions. *American Naturalist*, **140**, 1–29.

Ives, A. R. (1992b). Density-dependent and density-independent aggregation in model host–parasitoid systems. *American Naturalist*, **140**, 912–937.

Ives, W. G. H. (1976). The dynamics of larch sawfly (Hymenoptera: Tenthredinidae) populations in southeastern Manitoba. *Canadian Entomologist*, **108**, 701–730.

Ivlev, V. S. (1961). *Experimental ecology of the feeding of fish* (transl. from Russian by D. Scott). Yale University Press, New Haven, CT.

Iwao, S. (1971). Dynamics of numbers of a phytophagous lady-beetle, *Epilachna vigintioctomaculata*, living in patchily distributed habitats. In *Dynamics of populations* (ed. P. J. Den Boer and G. R. Gradwell), pp. 129–147. Centre for Agricultural Publishing and Documentation, Wageningen.

Iwasa, Y., Higashi, M., and Yamamura, N. (1981). Prey distribution as a factor determining the choice of optimal foraging strategy. *American Naturalist*, **117**, 710–723.

Jeffries, M. J. and Lawton, J. H. (1984). Enemy free space and the structure of ecological communities. *Biological Journal of the Linnean Society*, **23**, 269–286.

Jones, T. H. and Hassell, M. P. (1988). Patterns of parasitism by *Trybliographa rapae*, a cynipid parasitoid of the cabbage root fly, under laboratory and field conditions. *Ecological Entomology*, **13**, 309–317.

Jones, T. H., Hassell, M. P., and Pacala, S. W. (1993). Spatial heterogeneity and the population dynamics of a host–parasitoid system. *Journal of Animal Ecology*, **62**, 251–262.

Jones, T. H., Godfray, H. C. J., and Hassell, M. P. (1996). Relative movement patterns of a tephritid fly and its parasitoid wasps. *Oecologia*, **106**, 317–324.

Juliano, S. A. (1993). Nonlinear curve fitting: predation and functional response curves. In *Design and analysis of ecological experiments* (ed. S. M. Scheiner and J. Gurevich), pp. 159–168. Chapman and Hall, New York.

Juliano, S. A. and Williams, F. M. (1987). A comparison of methods for estimating the functional response parameters of the random predator equation. *Journal of Animal Ecology*, **56**, 641–653.

Kaitala, V., Ylikarjula, J., and Heino, M. (1999). Dynamic complexities in host–parasitoid interaction. *Journal of Theoretical Biology*, **197**, 331–341.

Kakehashi, N., Suzuki, Y., and Iwasa, Y. (1984). Niche overlap of parasitoids in host–parasitoid systems: its consequence to single versus multiple introduction controversy in biological control. *Journal of Applied Ecology*, **21**, 115–131.

Kaneko, K. (1992). Overview of coupled map lattices. *Chaos*, **2**, 279–282.

Kaneko, K. (1993). The coupled map lattice: introduction, phenomenology, Lyapunov analysis, thermodynamics and applications. In *Theory and applications of coupled map lattices* (ed. K. Kaneko), pp. 1–49. Wiley, Chichester.

Kareiva, P. M. (1989). Renewing the dialogue between theory and experiments in population ecology. In *Perspectives in ecological theory* (ed. J. Roughgarden, R. M. May, and S. A. Levin), pp. 68–88. Princeton University Press, Princeton, NJ.

Kareiva, P. (1990). Population dynamics in spatially complex environments: theory and data. *Proceedings of the Royal Society of London, Series B*, **330**, 175–190.

Kato, M. (1994). Alternation of bottom-up and top-down regulation in a natural population of an agromyzid leafminer, *Chromatomyia suikazurae*. *Oecologia*, **97**, 9–16.

Kazmer, D. J. and Luck, R. F. (1995). Field tests of the size fitness hypothesis in the egg parasitoid *Trichogramma praetiosum*. *Ecology*, **76**, 412–425.

Kermack, W. O. and McKendrick, A. G. (1927). A contribution to the mathematical theory of epidemics. *Proceedings of the Royal Society, Series A*, **115**, 700–721.

Kfir, R., Podoler, H., and Rosen, D. (1976). The area of discovery and searching strategy of a primary parasite and two hyperparasites. *Ecological Entomology*, **1**, 287–295.

Kidd, N. A. C., Smith, S. D. J., Lewis, G. B., and Carter, C. I. (1990). Interactions between host–plant chemistry and the population dynamics of conifer aphids. In *Population dynamics of forest insects* (ed. A. D. Watt, S. R. Leather, M. D. Hunter, and N. A. C. Kidd), pp. 183–193. Intercept, Andover.

Kingsland, S. E. (1996). Evolutionary theory and the foundations of population ecology: the work of A. J. Nicholson (1895–1969). In *Frontiers of population ecology* (ed. R. B. Floyd, A. W. Sheppard, and P. J. De Barro), pp. 13–25. CSIRO, Melbourne.

Kiritani, K. (1977). A systems approach to pest management of the green rice leafhopper. In *Proceedings of a Conference on Pest Management*, 25–29 October, 1976 (ed. G. A. Norton and C. S. Holling), pp. 229–252. International Institute for Applied Systems Analysis, Laxenburg, Austria.

Klomp, H. (1966). The dynamics of a field population of the pine looper, *Bupalus piniarius* (Lepidoptera: Geom.). *Advances in Ecological Research*, **3**, 207–305.

Klopfer, E. D. and Ives, A. R. (1997). Aggregation and the coexistence of competing parasitoid species. *Theoretical Population Biology*, **52**, 167–178.

Koebele, A. (1870). Report of a trip to Australia made under direction of the entomologist to investigate the natural enemies of the fluted scale. *Bulletin of the Bureau of Entomology, US Department of Agriculture*, **21**, 1–32.

Kot, M., Lewis, M. A., and van den Driessche, P. (1996). Dispersal data and the spread of invading organisms. *Ecology*, **77**, 2027–2042.

Kowalski, R. (1976). *Philonthus decorus* (Gr.) (Coleoptera: Staphylinidae): its biology in relation to its action as a predator of winter moth (*Operophtera brumata*) (Lepidoptera: Geometridae). *Pediobiologia*, **16**, 233–242.

Krebs, J. R. (1970). Regulation of numbers of the great tit (Aves: Passeriformes). *Journal of Zoology*, **162**, 317–333.

Krebs, J. R. and Kacelnik, A. (1991). Decision-making. In *Behavioural ecology, an evolutionary approach* (ed. J. R. Krebs and N. B. Davies), pp. 105–136. Blackwell Science, Oxford.

Krivan, V. (1996). Optimal foraging and predator–prey dynamics. *Theoretical Population Biology*, **49**, 265–290.

Krivan, V. (1997). Dynamic ideal free distribution: effects of optimal patch choice on predator–prey dynamics. *American Naturalist*, **149**, 164–178.

Krivan, V. (1998). Effects of optimal antipredator behavior of prey on predator–prey dynamics: the role of refuges. *Theoretical Population Biology*, **53**, 131–142.

Lane, S. D., Mills, N. J., and Getz, W. M. (1999). The effects of parasitoid fecundity and host taxon on the biological control of insect pests: the relationship between theory and data. *Ecological Entomology*, **24**, 181–190.

Laraichi, M. (1978). Etude de la competition intra- et interspecifique chez les parasites oophages des punaises des bles. *Entomophaga*, **23**, 115–120.

Laska, M. S. and Wootton, J. T. (1998). Theoretical concepts and empirical approaches to measuring interaction strength. *Ecology*, **79**, 461–476.

Lawrence, P. O. (1981). Interference competition and optimal host selection in the parasitic wasp *Biosteres longicaudatus*. *Annals of the Entomological Society of America*, **74**, 540–544.

Lawton, J. H. (1986). The effects of parasitoids on phytophagous insect communties. In *Insect parasitoids* (ed. J. K. Waage and D. J. Greathead), pp. 265–287. Academic Press, London.

Lawton, J. H. (1993). Range, population abundance and conservation. *Trends in Ecology and Evolution*, **8**, 409–413.

Lawton, J. H. and Hassell, M. P. (1984). Interspecific competition in insects. In *Ecological entomology* (ed. C. B. Huffaker and R. L. Rabb), pp. 451–495. Wiley, New York.

Lawton, J. H. and McNeill, S. (1979). Between the devil and the deep blue sea: on the problem of being a herbivore. In *Population Dynamics* (ed. R. M. Anderson, B. D. Turner, and L. R. Taylor), pp. 223–244. Blackwell Science, Oxford.

Lawton, J. H. and Pimm, S. L. (1978). Population dynamics and the length of food chains. *Nature, London*, **272**, 189–190.

Lawton, J. H. and Strong, D. R. (1981). Community patterns and competition in folivorous insects. *American Naturalist*, **118**, 317–338.

Lawton, J. H., Beddington, J. R., and Bonser, R. (1974). Switching in invertebrate predators. In *Ecological stability* (ed. M. B. Usher and M. H. Williamson), pp. 141–158. Chapman and Hall, London.

Lawton, J. H., Bignell, D. E., Bolton, B., Bloemers, G. F., Eggleton, P., Hammond, P. M., Hodda, M., Holt, R. D., Larsen, T. B., Mawdsley, N. A., and Stork, N. E. (1998). Biodiversity inventories, indicator taxa and effects of habitat modification in tropical forest. *Nature, London*, **391**, 72–76.

Lei, G. C. and Hanski, I. (1997). Metapopulation structure of *Cotesia melitaearum*, a specialist parasitoid of the butterfly *Melitaea cinxia*. *Oikos*, **78**, 91–100.

Lei, G. C. and Hanski, I. (1998). Spatial dynamics of two competing specialist parasitoids in a host metapopulation. *Journal of Animal Ecology*, **67**, 422–433.

Leirs, H., Stenseth, N. C., Nichols, J. D., Hines, J. E., Verhagen, R., and Verheyen, W.

(1997). Stochastic seasonality and nonlinear density-dependent factors regulate population size in an African rodent. *Nature, London*, **389**, 176–180.

Lessells, C. M. (1985). Parasitoid foraging: should parasitism be density dependent? *Journal of Animal Ecology*, **54**, 27–41.

Levin, S. A. (1992). The problem of pattern and scale in ecology. *Ecology*, **73**, 1943–1967.

Levin, S. A. (1994). Patchiness in marine and terrestrial systems: from individuals to populations. *Philosophical Transactions of the Royal Society, London B*, **343**, 99–103.

Levins, R. (1968). *Evolution in changing environments*. Princeton University Press, Princeton, NJ.

Levins, R. (1969). Some demographic and genetic consequences of environmental heterogeneity for biological control. *Bulletin of the Entomological Society of America*, **15**, 237–240.

Levins, R. (1970). Extinction. *Lecture Notes in Mathematics*, **2**, 75–107.

Levins, R. and Culver, D. (1971). Regional coexistence of species and competition between rare species. *Proceedings of the National Academy of Sciences USA*, **68**, 1246–1248.

Lewis, W. J. and Tumlinson, J. H. (1988). Host selection by chemically mediated associative learning in a parasitic wasp. *Nature, London*, **331**, 257–259.

Lewis, W. J. and Vinson, S. B. (1968). Immunological relationships between the parasite *Cardiochiles nigriceps* Viereck and certain *Heliothis* species. *Journal of Insect Physiology*, **14**, 613–626.

Liljesthrom, G. G. and Bernstein, C. (1992). Density dependence and regulation in the system *Nezara viridula* (L.). Hemiptera: Pentatomidae), host and *Trichopoda giacomellii* (Blanchard). Diptera: Tachinidae), parasitoid. *Oecologia*, **84**, 45–52.

Lotka, A. J. (1925). *Elements of physical biology*. Williams and Wilkins, Baltimore, MD.

Luck, R. F. (1990). Evalulation of natural enemies for biological control: a behavioural approach. *Trends in Ecology and Evolution*, **5**, 196–199.

Luck, R. F. and Podoler, H. (1985). Competitive exclusion of *Aphytis lingnanensis* by *A. melinus*: potential role of host size. *Ecology*, **66**, 904–913.

Luck, R. F., Podoler, H., and Kfir, R. (1982). Host selection and egg allocation behaviour of *Aphytis melinus* and *Aphytis lingnanensis*: a comparison of two facultatively gregarious parasitoids. *Ecological Entomology*, **7**, 397–408.

Ludwig, D., Jones, D. D., and Holling, C. S. (1978). Qualitative analysis of insect outbreak systems: the spruce budworm and forest. *Journal of Animal Ecology*, **47**, 315–332.

Lynch, L. D. (1998). Indirect mutual interference and the $CV^2 > 1$ rule. *Oikos*, **83**, 318–326.

Lynch, L. D., Bowers, R. G., Begon, M., and Thompson, D. J. (1998). A dynamic refuge model and population regulation by insect parasitoids. *Journal of Animal Ecology*, **67**, 270–279.

Lyons, L. A. (1964). The spatial distribution of two pine sawflies and methods of sampling for the study of population dynamics. *Canadian Entomologist*, **96**, 1373–1407.

MacArthur, R. H. (1968). The theory of the niche. In *Population biology and evolution* (ed. R. C. Lewontin), pp. 159–176. Syracuse University Press, Syracuse, NY.

MacArthur, R. H. and Levins, R. (1964). Competition, habitat selection, and character displacement in a patchy environment. *Proceedings of the National Academy of Sciences USA*, **51**, 1207–1210.

MacDonald, N. (1989). Biological delay systems: linear stability theory. Cambridge University Press, Cambridge.

Mackauer, M. (1990). Host discrimination and larval competition in solitary endoparasitoids. In *Critical issues in biological control* (eds. M. Mackauer, L. E. Ehler and J. Roland), pp. 41–62. Intercept, Andover.

Mackerras, I. (1970). Alexander John Nicholson. *Records of the Australian Academy of Science*, **2**, 66–81.

Manly, B. F. J. (1990). *Stage-structured populations. Sampling, analysis and simulation.* Chapman and Hall, London.

Mariau, D. and Morin, J. P. (1972). La biologie de *Coelaenomedera elaeidis*. IV. La dynamique des populations du ravageur et de ses parasites. *Oleagineaux*, **27**, 469–474.

Marrow, P., Law, R., and Cannings, C. (1992). The coevolution of predator–prey interactions: ESSs and Red Queen dynamics. *Proceedings of the Royal Society of London, B*, **250**, 133–141.

Maton, J. L., and Harrison, S. (1997). Spatial pattern formation in an insect host–parasitoid system. *Science*, **278**, 1619–1621.

May, R. M. (1973). On the relationship between various types of population models. *American Naturalist*, **107**, 46–57.

May, R. M. (1974*a*). Biological populations with non-overlapping generations: stable points, stable cycles and chaos. *Science*, **186**, 645–647.

May, R. M. (1974*b*). *Stability and complexity in model ecosystems*. Princeton University Press, Princeton, NJ.

May, R. M. (1977*a*). Predators that switch. *Nature, London*, **269**, 103–104.

May, R. M. (1977*b*). Thresholds and breakpoints in ecosystems with a multiplicity of stable states. *Nature, London*, **269**, 471–477.

May, R. M. (1978). Host–parasitoid systems in patchy environments: a phenomenological model. *Journal of Animal Ecology*, **47**, 833–843.

May, R. M. (1985). Regulation of populations with nonoverlapping generations by microparasites: a purely chaotic system. *American Naturalist*, **125**, 573–584.

May, R. M. (1994). The effects of spatial scale on ecological questions and answers. In *Large-scale ecology and conservation biology* (ed. P. J. Edwards, R. M. May, and N. R. Webb), pp. 1–17. Blackwell, Oxford.

May, R. M. and Anderson, R. M. (1978). Regulation and stability of host–parasite population interactions. II. Destabilizing processes. *Journal of Animal Ecology*, **47**, 249–267.

May, R. M. and Hassell, M. P. (1981). The dynamics of multiparasitoid–host interactions. *American Naturalist*, **117**, 234–261.

May, R. M. and Hassell, M. P. (1988). Population dynamics and biological control. *Philosophical Transactions of the Royal Society, London, Series B*, **318**, 129–169.

May, R. M. and Oster, G. F. (1976). Bifurcations and dynamic complexity in simple ecological models. *American Naturalist*, **110**, 573–600.

May, R. M., Hassell, M. P., Anderson, R. M., and Tonkyn, D. W. (1981). Density dependence in host–parasitoid models. *Journal of Animal Ecology*, **50**, 855–865.

Maynard Smith, J. (1974). *Models in ecology*. Cambridge University Press, Cambridge.

Maynard Smith, J. and Slatkin, M. (1973). The stability of predator–prey systems. *Ecology*, **54**, 384–391.

McMurtrie, R. (1978). Persistence and stability of single-species and predator–prey systems in spatially heterogeneous environments. *Mathematical Biosciences*, **39**, 11–51.

McNair, J. N. (1986). The effects of refuges on predator–prey interactions: a reconsideration. *Theoretical Population Biology*, **29**, 38–63.

McNeill, S. N. (1973). The dynamics of a population of *Leptoterna dolabrata* (Heteroptera:Miridae) in relation to its food resources. *Journal of Animal Ecology*, **42**, 495–507.

Memmott, J. and Godfray, H. C. J. (1993). Parasitoid webs. In *Hymenoptera and biodiversity* (ed. J. LaSalle and I. D. Gauld), pp. 217–234. CAB International, Wallingford.

Memmott, J., Godfray, H. C. J., and Bolton, B. (1993). Predation and parasitism in a tropical herbivore community. *Ecological Entomology*, **18**, 348–352.

Memmott, J., Godfray, H. C. J., and Gauld, I. D. (1994). The structure of a tropical host–parasitoid community. *Journal of Animal Ecology*, **63**, 521–540.

Messenger, P. S. and van den Bosch, R. (1971). The adaptability of introduced biological control agents. In *Biological Control* (ed. C. B. Huffaker), pp. 68–92. Plenum, New York.

Mills, N. J. (1991). Searching strategies and attack rates of parasitoids of the ash bark beetle (*Leperisinus varius*) and its relevance to biological control. *Ecological Entomology*, **16**, 461–470.

Miyashita, K. (1955). Some considerations on the population fluctuation of the rice stem borer. *Bulletin of the National Institute of Agricultural Sciences, Japan, Series C*, **5**, 99–106.

Mohamed, M. A. and Coppel, H. C. (1986). Sex-ratio regulation in *Brachymeria intermedia*, a pupal gypsy-moth parasitoid. *Canadian Journal of Zoology*, **64**, 1412–1415.

Montgomery, M. E. and Wallner, W. E. (1988). The gypsy moth, a westward migrant. In *Dynamics of forest insect populations: patterns, causes, implications* (ed. A. A. Berryman), pp. 353–376. Plenum, New York.

Moody, A. L. and Ruxton, G. D. (1996). The intensity of interference varies with feed density: support for behaviour-based models of interference. *Oecologia*, **108**, 446–449.

Moran, P. A. P. (1950). Some remarks on animal population dynamics. *Biometriks*, **6**, 250–258.

Mori, H. and Chant, D. A. (1966). The influence of prey density, relative humidity, and starvation on the predaceous behaviour of *Phytoseiulus persimilis* Athias-Henriot (Acarina: Phytoseiidae). *Canadian Journal of Zoology*, **44**, 483–491.

Morris, R. F. (1959). Single-factor analysis in population dynamics. *Ecology*, **40**, 580–588.

Muesebeck, C. F. W. (1931). *Monodontomerus aereus* Walker, both a primary and secondary parasite of the brown-tail moth and gypsy moth. *Journal of Agricultural Research*, **43**, 445–460.

Müller, C. B. and Godfray, H. C. J. (1999). Indirect interactions in aphid–parasitoid communities. *Researches on Population Ecology*, **41**, 93–106.

Müller, C. B., Adriaanse, I. C. T., Belshaw, R., and Godfray, H. C. J. (1999). The structure of an aphid–parasitoid community. *Journal of Animal Ecology*, **68**, 346–370.

Munster-Swendsen, M. (1980). The distribution in time and space of parasitism in *Epinotia tedella* (Cl.). Lepidoptera: Tortricidae). *Ecological Entomology*, **5**, 373–383.

Munster-Swendsen, M. (1985). A simulation study of primary-, clepto- and hyper-parasitism. *Journal of Animal Ecology*, **54**, 683–695.

Munster-Swendsen, M. and Nachman, G. (1978). Asynchrony in insect host–parasite interaction and its effect on stability, studied by a simulation model. *Journal of Animal Ecology*, **47**, 159–171.

Murdoch, W. W. (1966). Community structure, population control and competition—a critique. *American Naturalist*, **100**, 219–226.

Murdoch, W. W. (1969). Switching in general predators: experiments on predator specificity and stability of prey populations. *Ecological Monographs*, **39**, 335–354.

Murdoch, W. W. and Oaten, A. (1975). Predation and population stability. *Advances in Ecological Research*, **9**, 1–131.

Murdoch, W. W. and Stewart-Oaten, A. (1989). Aggregation by parasitoids and predators: effects on equilibrium and stability. *American Naturalist*, **134**, 288–310.

Murdoch, W. W., Avery, S., and Smyth, M. E. B. (1975). Switching in predatory fish. *Ecology*, **56**, 1094–1105.

Murdoch, W. W., Reeve, J. D., Huffaker, C. B., and Kennett, C. E. (1984*a*). Biological control of olive scale and its relevance to ecological theory. *American Naturalist*, **123**, 371–392.

Murdoch, W. W., Scott, M. A., and Ebsworth, P. (1984*b*). Effects of the general predator, *Notonecta* (Hemiptera) upon a fresh-water community. *Journal of Animal Ecology*, **53**, 791–808.

Murdoch, W. W., Nisbet, R. M., Blythe, S. P., Gurney, W. S. C., and Reeve, J. D. (1987). An invulnerable age class and stability in delay-differential parasitoid-host models. *American Naturalist*, **129**, 263–282.

Murdoch, W. W., Luck, R. F., Walde, S., Reeve, J. D., and Yu, D. S. (1989). A refuge for red scale under control by *Aphytis*: structural aspects. *Ecology*, **70**, 1707–1714.

Murdoch, W. W., Briggs, C. J., Nisbet, R. M., Gurney, W. S. C., and Stewart-Oaten, A. (1992*a*). Aggregation and stability in metapopulation models. *American Naturalist*, **140**, 41–58.

Murdoch, W. W., Nisbet, R. M., Luck, R. F., Godfray, H. C. J., and Gurney, W. S. C. (1992*b*). Size-selective sex-allocation and host feeding in a parasitoid–host model. *Journal of Animal Ecology*, **61**, 533–541.

Murdoch, W. W., Luck, R. F., Swarbrick, S. L., Walde, S., Yu, D. S., and Reeve, J. D. (1995). Regulation of an insect population under biological control. *Ecology*, **76**, 206–217.

Murdoch, W. W., Briggs, C. J., and Nisbet, R. M. (1996*a*). Competitive displacement and biological control in parasitoids—a model. *American Naturalist*, **148**, 807–826.

Murdoch, W. W., Swarbrick, S. L., Luck, R. F., Walde, S., and Yu, D. S. (1996*b*). Refuge dynamics and metapopulation dynamics—an experimental test. *American Naturalist*, **147**, 424–444.

Murdoch, W. W., Briggs, C. J., and Nisbet, R. M. (1997). Dynamical effects of host size- and parasitoid state-dependent attacks by parasitoids. *Journal of Animal Ecology*, **66**, 542–556.

Naeem, S. and Fenchel, T. (1994). Population growth on a patchy resource—some insights provided by studies of a histophagous protozoan. *Journal of Animal Ecology*, **63**, 399–409.

Nakamura, K. and Ohgushi, T. (1981). Studies on the population dynamics of a thistle-feeding lady beetle, *Henosepilachna pustulosa* (Kuno) in a cool temperate climax forest. II. Lifetables, key factor analysis, and detection of regulatory mechanisms. *Researches on Population Ecology*, **23**, 210–231.

Nappi, A. J. (1975). Parasite encapsulation in insects. In *Invertebrate immunity* (ed. K. Maramorosch and R. E. Shope), pp. 293–326. Academic Press, New York.

Nee, S. and May, R. M. (1992). Dynamics of metapopulations: habitat destruction and competitive coexistence. *Journal of Animal Ecology*, **61**, 37–40.

Nee, S., May, R. M., and Hassell, M. P. (1997). Two-species metapopulation models. In *Metapopulation biology. Ecology*, Genetics, and Evolution (ed. I. Hanski and M. E. Gilpin), pp. 123–147. Academic Press, San Diego, CA.

Nicholson, A. J. (1933). The balance of animal populations. *Journal of Animal Ecology*, **2**, 131–178.

Nicholson, A. J. (1947). Fluctuation of animal populations. *Proceedings of the Australian and New Zealand Association for the Advancement of Science*, **26**, 1–14.

Nicholson, A. J. (1950). Population oscillations caused by competition for food. *Nature, London*, **165**, 476–477.

Nicholson, A. J. (1954). An outline of the dynamics of animal populations. *Australian Journal of Zoology*, **2**, 9–65.

Nicholson, A. J. (1957). The self-adjustment of populations to change. *Cold Spring Harbor Symposium on Quantitative Biology*, **22**, 153–173.

Nicholson, A. J. and Bailey, V. A. (1935). The balance of animal populations. Part 1. *Proceedings of the Zoological Society of London*, **3**, 551–598.

Nieminen, M. and Hanski, I. (1998). Metapopulations of moths on islands: a test of two contrasting models. *Journal of Animal Ecology*, **67**, 149–160.

Nisbet, R. M. and Gurney, W. S. C. (1982). Modelling fluctuating populations. Wiley, New York.

Nisbet, R. M. and Gurney, W. S. C. (1983). The systematic formulation of population models for insects with dynamically varying instar duration. *Theoretical Population Biology*, **23**, 114–135.

Nowak, M. A., Bonhoeffer, S., and May, R. M. (1994). More spatial games. *International Journal of Bifurcation and Chaos*, **4**, 33–56.

Pacala, S. W. (1997). Dynamics of plant communities. In *Plant ecology* (ed. M. J. Crawley), pp. 532–555. Blackwell Science, Oxford.

Pacala, S. W. and Hassell, M. P. (1991). The persistence of host–parasitoid associations in patchy environments. II. Evaluation of field data. *American Naturalist*, **138**, 584–605.

Pacala, S. W. and Silander, J. A. (1990). Field-tests of neighborhood population-dynamic models of two annual weed species. *Ecological Monographs*, **60**, 113–134.

Pacala, S. W. and Tilman, D. (1994). Limiting similarity in mechanistic and spatial models of plant competition in heterogeneous environments. *American Naturalist*, **143**, 222–257.

Pacala, S. W., Hassell, M. P., and May, R. M. (1990). Host–parasitoid associations in patchy environments. *Nature, London*, **344**, 150–153.

Pacala, S. W., Canham, C. D., and Silander, J. A. (1993). Forest models defined by field measurements. 1. The design of a northeastern forest simulator. *Canadian Journal of Forest Research*, **23**, 1980–1988.

Pacala, S. W., Canham, C. D., Saponara, J., Silander, J. A., Kobe, R. K., and Ribbens, E. (1996). Forest models defined by field measurements: estimation, error analysis and dynamics. *Ecological Monographs*, **66**, 1–43.

Paine, R. T. (1966). Food web complexity and species diversity. *American Naturalist*, **100**, 65–75.

Paine, R. T. (1974). Intertidal community structure. *Oecologia*, **15**, 93–120.

Paine, R. T. (1984). Ecological determinism in the competition for space. *Ecology*, **65**, 1339–1348.

Papaj, D. R. and Vet, L. E. M. (1990). Odor learning and foraging success in the parasitoid, *Leptopilina heterotoma*. *Journal of Chemical Ecology*, **16**, 3137–3150.

Pemberton, C. E. and Willard, H. F. (1918). Interactions of fruit-fly parasites in Hawaii. *Journal of Agricultural Research*, **12**, 551–598.

Pielou, E. C. (1969). *An introduction to mathematical ecology*. Wiley, New York.

Pimm, S. L. and Lawton, J. H. (1977). Numbers of trophic levels in ecological communities. *Nature, London*, **268**, 329–331.

Pimm, S. L. and Lawton, J. H. (1978). On feeding on more than one trophic level. *Nature, London*, **275**, 542–544.

Pimm, S. L. and Lawton, J. H. (1980). Are food webs divided into compartments? *Journal of Animal Ecology*, **49**, 879–898.

Pimm, S. L., Jones, H. L., and Diamond, J. (1988). On the risk of extinction. *American Naturalist*, **132**, 757–785.

Podoler, H. and Rogers, D. J. (1975). A new method for the identification of key factors from life-table data. *Journal of Animal Ecology*, **44**, 85–115.

Prestidge, R. and McNeill, S. N. (1983). The role of nitrogen in the ecology of grassland Auchenorrhyncha. In *Nitrogen as an ecological factor* (ed. J. A. Lee, S. McNeill, and I. H. Rorison), pp. 257–281. Blackwell Science, Oxford.

Price, P. W. (1970). Trail odours: recognition by insects parasitic in cocoons. *Science*, **170**, 546–547.

Price, P. W. (1972). Behaviour of the parasitioid *Pleolophus indistinctus* (Hymenoptera: Ichneumonidae). *Annals of the Entomological Society of America*, **63**, 1502–1509.

Price, P. W., Fernandes, G. W., and Declearck-Floate, R. (1995). Gall-inducing insect herbivores in multitrophic systems. In *Multitrophic interactions in terrestrial systems* (ed. A. C. Gange and V. K. Brown), pp. 239–255. Blackwell Science, Oxford.

Pschorn-Walcher, H. and Zwölfer, H. (1968). Konkurrenzerscheinungen in Parasitenkomplexen als Problem der biologischen Schädlingsbekampfung. *Anzeiger für Schadlingskunde*, **41**, 71–76.

Rabinovich, J. E. (1984). Chagas' disease: modelling transmission and control. In *Pest and pathogen control strategic, tactical and policy models* (ed. G. R. Conway), pp. 58–72. Wiley, New York.

Rand, D. A. and Wilson, H. B. (1995). Using spatio-temporal chaos and intermediate-scale determinism to quantify spatially extended ecosystems. *Proceedings of the Royal Society of London, B*, **259**, 111–117.

Rand, D. A. and Wilson, H. B. (1991). Chaotic stochasticity: a ubiquitous source of unpredictability in epidemics. *Proceedings of the Royal Society of London, B*, **246**, 179–184.

Rand, D. A., Keeling, M., and Wilson, H. B. (1995). Invasion, stability and evolution to criticality in spatially extended, artificial host-pathogen ecologies. *Proceedings of the Royal Society of London, B*, **259**, 55–63.

Redfern, M., Jones, T. H., and Hassell, M. P. (1992). Heterogeneity and density dependence in a field study of a tephritid–parasitoid interaction. *Ecological Entomology*, **17**, 255–262.

Rees, M. and Paynter, Q. (1997). Biological control of Scotch broom: modelling the determinants of abundance and the potential impact of introduced insect herbivores. *Journal of Applied Ecology*, **34**, 1203–1221.

Rees, M., Grubb, P. J., and Kelly, D. (1996). Quantifying the impact of competition and spatial heterogeneity on the structure and dynamics of a four-species guild of winter annuals. *American Naturalist*, **147**, 1–32.

Rees, M., Sheppard, A., Briese, D., and Mangel, M. (2000). Evolution of size-dependent flowering in *Onopordum illyricim*: a quantitative assessment of the role of stochastic selection pressures. *American Naturalist* (In press.)

Reeve, J. D. (1988). Environmental variability, migration, and persistence in host–parasitoid systems. *American Naturalist*, **132**, 810–836.

Reeve, J. D. (1990). Stability, variability, and persistence in host–parasitoid systems. *Ecology*, **71**, 422–426.

Reeve, J. D. and Murdoch, W. W. (1985). Aggregation by parasitoids in the successful control of the California red scale: a test of theory. *Journal of Animal Ecology*, **54**, 797–816.

Reeve, J. D. and Murdoch, W. W. (1986). Biological control of the parasitoid *Aphytis melinus*, and population stability of the California red scale. *Journal of Animal Ecology*, **55**, 1069–1082.

Reeve, J. D., Kerans, B. L., and Chesson, P. L. (1989). Combining different forms of parasitoid aggregation: effects on stability and patterns of parasitism. *Oikos*, **56**, 233–239.

Reeve, J. D., Cronin, J. T., and Strong, D. R. (1994*a*). Parasitism and generation cycles in a salt-marsh planthopper. *Journal of Animal Ecology*, **63**, 912–920.

Reeve, J. D., Cronin, J. T., and Strong, D. R. (1994*b*). Parasitoid aggregation and the stabilization of a salt-marsh host–parasitoid system. *Ecology*, **75**, 288–295.

Rejmanek, M. and Stary, P. (1979). Connectance in real biotic communities and critical values for stability of model ecosystems. *Nature, London*, **280**, 311–313.

Rhoades, D. F. (1985). Offensive–defensive interactions between herbivores and plants—their relevance in herbivore population-dynamics and ecological theory. *American Naturalist*, **125**, 205–238.

Ricker, W. E. (1954). Stock and recruitment. *Journal of the Fish Research Board of Canada*, **11**, 559–623.

Rigler, F. H. (1961). The relation between concentration of food and feeding rate of *Daphnia magna* Straus. *Canadian Journal of Zoology*, **39**, 857–868.

Rizki, R. M. and Rizki, T. M. (1984). Selective destruction of a host blood cell type by a parasitoid wasp. *Proceedings of the National Academy of Sciences USA*, **81**, 6154–6158.

Rogers, D. J. (1972). Random searching and insect population models. *Journal of Animal Ecology*, **41**, 369–383.

Rogers, D. J. and Hassell, M. P. (1974). General models for insect parasite and predator searching behaviour: interference. *Journal of Animal Ecology*, **43**, 239–253.

Rohani, P., and Miramontes, O. (1995). Host–parasitoid metapopulations—the consequences of parasitoid aggregation on spatial dynamics and searching efficiency. *Proceedings of the Royal Society of London, B*, **260**, 335–342.

Rohani, P., Godfray, H. C. J., and Hassell, M. P. (1994*a*). Aggregation and the dynamics of host–parasitoid systems: a discrete-generation model with within-generation redistribution. *American Naturalist*, **144**, 491–509.

Rohani, P., Miramontes, O., and Hassell, M. P. (1994*b*). Quasiperiodicity and chaos in population models. *Philosophical Transactions of the Royal Society, London, Series B*, **258**, 17–22.

Rohani, P., May, R. M., and Hassell, M. P. (1996). Metapopulations and equilibrium stability: the effects of spatial structure. *Journal of Theoretical Biology*, **181**, 97–109.

Roininen, H., Price, P. W., and Tahvanainen, J. (1996). Bottom-up and top-down influences in the trophic system of a willow, a galling sawfly, parasitoids and inquilines. *Oikos*, **77**, 44–50.

Roitberg, B. D. and Prokopy, R. J. (1987). Insects that mark host plants. *BioScience*, **37**, 400–406.

Roland, J. (1986). Parasitism of winter moth in British Columbia during build-up of its parasitoid *Cyzenis albicans*: attack rate on oak v. apple. *Journal of Animal Ecology*, **55**, 215–234.

Roland, J. (1988). Decline in winter moth populations in North America: direct versus indirect effect of introduced parasites. *Journal of Animal Ecology*, **57**, 523–531.

Roland, J. (1989). Success and failure of *Cyzenis albicans* in controlling its host the winter moth. PhD Thesis, University of British Columbia, Vancouver, Canada.

Roland, J. (1994). After the decline—what maintains low winter moth density after successful biological control. *Journal of Animal Ecology*, **63**, 392–398.

Rotheram, S. M. (1967). Immune surface of eggs of a parasitic insect. *Nature, London*, **214**, 700.

Rott, A. S., Müller, C. B., and Godfray, H. C. J. (1998). Indirect population interaction between two aphid species. *Ecology Letters*, **1**, 99–103.

Roughgarden, J. D. and Feldman, M. (1975). Species packing and predation pressure. *Ecology*, **56**, 489–492.

Royama, T. (1970). Factors governing the hunting behaviour and selection of food by the great tit (*Parus major* L.). *Journal of Animal Ecology*, **39**, 619–669.

Royama, T. (1971). A comparative study of models for predation and parasitism. *Researches on Population Ecology*, **1**, 1–91.

Ruggiero, A., Lawton, J. H., and Blackburn, T. M. (1998). The geographic ranges of mammalian species in South America: spatial patterns in environmental resistance and anisotropy. *Journal of Biogeography*, **25**, 1093–1103.

Ruxton, G. D. and Rohani, P. (1996). The consequences of stochasticity for self-organized spatial dynamics, persistence and coexistence in spatially extended host–parasitoid communities. *Proceedings of the Royal Society of London, Series B*, **263**, 625–631.

Ruxton, G. D., Gurney, W. S. C., and de Roos, A. M. (1992). Interference and generation cycles. *Theoretical Population Biology*, **42**, 235–253.

Sabelis, M. W., Diekmann, O., and Jansen, V. A. A. (1991). Metapopulation persistence despite local extinction: predator–prey patch models of the Lotka–Volterra type. *Biological Journal of the Linnean Society*, **42**, 267–283.

Saccheri, I., Kuussaari, M., Kankare, M., Vikman, P., Fortelius, W., and Hanski, I.

(1998). Inbreeding and extinction in a butterfly metapopulation. *Nature, London*, **392**, 491–494.

Salt, G. (1963). The defence reactions of insects to metazoan parasites. *Parasitology*, **53**, 527–642.

Salt, G. (1968). The resistance of insect parasitoids to the defence reactions of their hosts. *Biological Reviews*, **43**, 200–232.

Salt, G. (1970). *The cellular defence reactions of insects*. Cambridge University Press, Cambridge.

Sasaba, T., Kiritani, K., and Urabe, T. (1973). A preliminary model to simulate the effect of insecticides on a spider-leafhopper system in the paddy field. *Researches on Population Ecology*, **15**, 9–22.

Sasaki, A. and Godfray, H. C. J. (1999). A model for the coevolution of resistance and virulence in coupled host–parasitoid interactions. *Proceedings of the Royal Society of London, Series B*, **266**, 455–463.

Schoener, T. W. (1991). Extinction and the nature of the metapopulation. *Acta Oecologia*, **12**, 53–75.

Schoener, T. W. and Spiller, D. A. (1987a). Effects of lizards on spider populations: manipulative reconstruction of a natural experiment. *Science*, **236**, 949–952.

Schoener, T. W. and Spiller, D. A. (1987b). High population persistence in a system with high turnover. *Nature, London*, **330**, 474–477.

Schonrogge, K., Stone, G. N., and Crawley, M. J. (1996). Alien herbivores and native parasitoids: rapid developments and structure of the parasitoid and inquiline complex in an invading gall wasp *Andricus quercuscalicis* (Hymenoptera: Cynipidae). *Ecological Entomology*, **21**, 71–80.

Schonrogge, K., Walker, P., and Crawley, M. J. (1999). Complex life cycles in *Andricus kollari* (Hymenoptera, Cynipidae) and their impact on associated parasitoid and inquiline species. *Oikos*, **84**, 293–301.

Schröder, D. (1974). A study of the interactions of the internal parasites of *Rhyacionia buoliana* (Lepidoptera: Olethreutidae). *Entomophaga*, **19**, 145–171.

Schwerdtfeger, F. (1935). Studien über den Mansenweschel einiger Forstschadhing. *Zeitschrift für Forst-u. Jagdwesen*, **67**, 15–38.

Settle, W. H. and Wilson, L. T. (1990a). Behavioral factors affecting differential parasitism by *Anagrus epos* (Hymenoptera, Mymaridae), of 2 species of erythroneuran leafhoppers (Homoptera, Cicadellidae). *Journal of Animal Ecology*, **59**, 877–891.

Settle, W. H. and Wilson, L. T. (1990b). Invasion by the variegated leafhopper and biotic interactions—parasitism, competition, and apparent competition. *Ecology*, **71**, 1461–1470.

Sheehan, W. and Hawkins, B. A. (1991). Attack strategy as an indicator of host range in metopiine and pimpline Ichneumonidae (Hymenoptera). *Ecological Entomology*, **16**, 129–131.

Shimada, M. (1999). Population fluctuation and persistence of one-host–two-parasitoid systems depending on resource distribution: from parasitizing behavior to population dynamics. *Researches on Population Ecology*, **41**, 69–79.

Shimada, M. and Tuda, M. (1996). Delayed density dependence and oscillatory population dynamics in overlapping-generation systems of a seed beetle *Callosobruchus chinensis*: matrix population model. *Oecologia*, **105**, 116–125.

Shorrocks, B. and Bingley, M. (1994). Priority effects and species coexistence—experiments with fungal-breeding *Drosophila*. *Journal of Animal Ecology*, **63**, 799–806.

Sibly, R. M. and Smith, R. H. (1998). Identifying key factors using lambda contribution analysis. *Journal of Animal Ecology*, **67**, 17–24.

Sih, A. (1981). Stability, prey density and age-dependent interference in an aquatic insect predator, *Notonecta hoffmani*. *Journal of Animal Ecology*, **50**, 625–636.

Sih, A., Crowley, P., McPeek, M., Petranka, J., and Strohmeier, K. (1985). Predation, competition, and prey communities—a review of field experiments. *Annual Review of Ecology and Systematics*, **16**, 269–311.

Silvertown, J. W. (1982). *Introduction to plant population ecology*. Longman, London.

Sinclair, A. R. E. (1973). Regulation, and population models for a tropical ruminant. *East African Wildlife*, **11**, 307–316.

Sinclair, A. R. E. (1989). Population regulation in animals. In *Ecological concepts: the contribution of ecology to an understanding of the natural world* (ed. J. M. Cherret), pp. 197–241. Blackwell Science, Oxford.

Skellam, J. G. (1951). Random dispersal in theoretical populations. *Biometrika*, **38**, 196–218.

Slobodkin, L. B. (1992). A summary of the special feature and comments on its theoretical context and importance. *Ecology*, **73**, 1564–1566.

Smith, H. S. (1929). Multiple parasitism: its relation to the biological control of insect pests. *Bulletin of Entomological Research*, **20**, 141–149.

Smith, H. S. (1935). The role of biotic factors in the determination of population densities. *Journal of Economic Entomology*, **28**, 873–898.

Smith, R. H. and Mead, R. (1974). Age structure and stability in models of prey–predator systems. *Theoretical Population Biology*, **6**, 308–322.

Soberon, J. (1986). The relationship between use and suitability of resources and its consequences to insect population size. *American Naturalist*, **127**, 338–357.

Solé, R. V. and Valls, J. (1992). Spiral waves, chaos and multiple attractors in lattice models of interacting populations. *Physics Letters A*, **166**, 123–128.

Solé, R. V., Bascompte, J., and Valls, J. (1992). Stability and complexity of spatially extended two-species competition. *Journal of Theoretical Biology*, **159**, 469–480.

Solomon, M. E. (1949). The natural control of animal populations. *Journal of Animal Ecology*, **18**, 1–35.

Southern, H. N. (1970). The natural control of a population of tawny owls (*Strix aluco*). *Journal of Zoology*, **162**, 197–285.

Southwood, T. R. E. (1976). *Ecological methods*. Chapman and Hall, London.

Southwood, T. R. E. and Comins, H. N. (1976). A synoptic population model. *Journal of Animal Ecology*, **45**, 949–965.

Southwood, T. R. E., Hassell, M. P., Reader, P. M., and Rogers, D. J. (1989). Population dynamics of the viburnum whitefly (*Aleurotrachelus jelinekii*). *Journal of Animal Ecology*, **58**, 921–942.

Spiller, D. A. and Schoener, T. W. (1990). Lizards reduce food consumption by spiders: mechanisms and consequences. *Oecologia*, **85**, 150–161.

Stephens, D. W. and Krebs, J. R. (1986). *Foraging theory*. Princeton University Press, Princeton, NJ.

Stiling, P. D. (1987). The frequency of density dependence in insect host–parasitoid systems. *Ecology*, **68**, 844–856.

Stiling, P. D. and Strong, D. R. (1982). Egg density and the intensity of parasitism in *Prokelisia marginata* (Homoptera: Delphacidae). *Ecology*, **63**, 1630–1635.

Stinner, R. E. and Lucas, H. L. (1976). Effects of contagious distributions of parasitoid eggs per host and of sampling vagaries on Nicholson's area of discovery. *Researches on Population Ecology*, **18**, 74–88.

Stoltz, D. B. and Vinson, S. B. (1979). Viruses and parasitism in insects. *Advances in Virus Research*, **24**, 125–170.

Strand, M. R. (1988). Variable sex ratio strategy of *Telenomus heliothidis* (Hymenoptera: Scelionidae): adaptation to host and conspecific density. *Oecologia*, **77**, 219–224.

Strong, D. R. (1986). Density-vague population change. *Trends in Ecology and Evolution*, **1**, 39–42.

Strong, D. R. (1989). Density independence in space and inconsistent temporal relationships for host mortality caused by a fairyfly parasitoid. *Journal of Animal Ecology*, **58**, 1065–1076.

Sugihara, G. and May, R. M. (1990). Nonlinear forecasting as a way of distinguishing chaos from measurement error in time series. *Nature, London*, **344**, 734–741.

Sutherland, W. J. (1983). Aggregation and the 'ideal free distribution'. *Journal of Animal Ecology*, **52**, 821–828.

Sutherland, W. J. (1996). *From individual behaviour to population ecology*. Oxford University Press, Oxford.

Swinton, J. and Anderson, R. M. (1995). Model frameworks for plant–pathogen interactions. In *Ecology of infectious diseases in natural populations* (ed. B. Grenfell and A. Dobson), pp. 280–294. Cambridge University Press, Cambridge.

Takahashi, F. (1968). Functional response to host density in a parasitic wasp, with reference to population regulation. *Researches on Population Ecology*, **10**, 54–68.

Taylor, A. D. (1988*a*). Host effects on larval competition in the gregarious parasitoid *Bracon hebetor*. *Journal of Animal Ecology*, **57**, 163–172.

Taylor, A. D. (1988*b*). Large-scale spatial structure and population dynamics in arthropod predator–prey systems. *Annales Zoologici Fennici*, **25**, 63–74.

Taylor, A. D. (1988*c*). Parasitoid competition and the dynamics of host–parasitoid models. *American Naturalist*, **132**, 417–436.

Taylor, A. D. (1990). Metapopulations, dispersal, and predator–prey dynamics: an overview. *Ecology*, **71**, 429–433.

Taylor, A. D. (1991). Studying metapopulation effects in predator–prey systems. *Biological Journal of the Linnean Society*, **42**, 305–323.

Taylor, A. D. (1993*a*). Aggregation, competition, and host–parasitoid dynamics: stability conditions don't tell it all. *American Naturalist*, **141**, 501–506.

Taylor, A. D. (1993*b*). Heterogeneity in host–parasitoid interactions: 'aggregation of risk' and the '$CV^2 > 1$' rule. *Trends in Ecology and Evolution*, **8**, 400–405.

Taylor, A. D. (1998). Environmental variability and the persistence of parasitoid–host metapopulation models. *Theoretical Population Biology*, **53**, 98–107.

Taylor, L. R., Woiwod, I. P., and Perry, J. N. (1979). The negative binomial as an ecological model and the density dependence of k. *Journal of Animal Ecology*, **48**, 289–304.

Teramoto, E., Kawasaki, K., and Shigesada. P. (1979). Switching effect of predation on competitive prey species. *Journal of Theoretical Biology*, **79**, 303–315.

Thomas, C. D. and Hanski, I. (1977). Butterfly metapopulations. In *Metapopulation biology: ecology, genetics, and evolution* (ed. I. Hanski and M. E. Gilpin), pp. 359–386. Academic Press, San Diego, CA.

Thompson, W. R. (1924). La théorie Mathématique de l'action des parasites entomophages et le facteur du hasard. *Annales de la Faculté des Sciences de Marseille*, **2**, 69–89.

Thompson, W. R. (1930). The utility of mathematical methods in relation to work on biological control. *Annals of Applied Biology*, **17**, 641–648.

Thompson, W. R. (1939). Biological control and the theories of interaction of populations. *Parasitology*, **31**, 299–388.

Tilman, D. (1981). Tests of resource competition theory using four species of Lake Michigan algae. *Ecology*, **62**, 802–815.

Tilman, D. (1982). *Resource competition and community structure*. Princeton University Press, Princeton, NJ.

Tilman, D. (1994). Competition and biodiversity in spatially structured habitats. *Ecology*, **75**, 2–16.

Tilman, D. and Sterner, R. W. (1984). Invasions of equilibria—tests of resource competition using two species of algae. *Oecologia*, **61**, 197–200.

Tilman, D. and Wedin, D. (1991). Dynamics of nitrogen competition between successional grasses. *Ecology*, **72**, 1038–1049.

Tilman, D., May, R. M., Lehman, C. L., and Nowak, M. A. (1994). Habitat destruction and the extinction debt. *Nature, London*, **371**, 65–66.

Tilman, D., Lehman, C. L., and Yin, C. J. (1997). Habitat destruction, dispersal, and deterministic extinction in competitive communities. *American Naturalist*, **149**, 407–435.

Tinbergen, L. (1960). The natural control of insects in pinewoods. 1: Factors influencing the intensity of predation by songbirds. *Archives Neelandaises de Zoologie*, **13**, 266–336.

Tostawaryk, W. (1972). The effect of prey defence on the functional response of *Podisus modestus* (Hemiptera: Pentatomidae) to densities of the sawflies *Neodiprion swainei* and *N. pratti banksianae* (Hymenoptera: Neodiprionidae). *Canadian Entomologist*, **104**, 61–69.

Townes, H. (1971). Ichneumonidae as biological control agents. *Proceedings of the Tall Timbers Conference on Ecological Animal Control by Habitat Management*, **3**, 235–248.

Trexler, J. C., McCulloch, C. E., and Travis, J. (1988). How can the functional response best be determined? *Oecologia*, **76**, 206–214.

Tuda, M. and Shimada, M. (1995). Developmental schedules and persistence of experimental host–parasitoid systems at two different temperatures. *Oecologia*, **103**, 283–291.

Turchin, P. (1990). Rarity of density dependence or population regulation with lags? *Nature, London*, **344**, 660–663.

Turchin, P. (1995). Population regulation: old arguments and a new synthesis. In *Population dynamics* (ed. N. Cappuccino and P. W. Price), pp. 19–40. Academic Press, New York.

Turchin, P. and Taylor, A. D. (1992). Complex dynamics in ecological time series. *Ecology*, **73**, 289–305.

Turlings, T. C. J., Tumlinson, J. H., and Lewis, W. J. (1990). Exploitation of herbivore-induced plant odours by host seeking parasitic wasps. *Science*, **250**, 1251–1253.

Turnbull, A. L. (1967). Population dynamics of exotic insects. *Bulletin of the Entomological Society of America*, **13**, 333–337.

Turnbull, A. L. and Chant, P. A. (1961). The practice and theory of biological control of insects in Canada. *Canadian Journal of Zoology*, **39**, 697–753.

Ueno, T. (1999). Host-size-dependent sex ratio in a parasitoid wasp. *Researches on Population Ecology*, **41**, 47–57.

Ullyett, G. C. (1949). Distribution of progeny by *Cryptus inornatus* Pratt (Hymenoptera: Ichneumonidae). *Canadian Entomologist*, **81**, 285–299.

Utida, S. (1941). Studies on experimental population of the azuki bean weevil *Callosobruchus chinensis* (L.). I. The effect of population density on the progeny populations. *Memoirs of the College of Agriculture, Kyoto University*, **48**, 1–31.

Utida, S. (1957). Cyclic fluctuations of population density intrinsic to the host parasite system. *Ecology*, **38**, 442–449.

van den Bosch, R. (1968). Comments on population dynamics of exotic insects. *Bulletin of the Entomological Society of America*, **14**, 112–115.

van Lenteren, J. C. and Bakker, K. (1976). Functional responses in invertebrates. *Netherlands Journal of Zoology*, **26**, 567–572.

van Valen, L. (1974). Predation and species diversity. *Journal of Theoretical Biology*, **44**, 19–21.

Varley, G. C. (1947). The natural control of population balance in the knapweed gall-fly (*Urophora jaceana*). *Journal of Animal Ecology*, **16**, 139–187.

Varley, G. C. and Gradwell, G. R. (1960). Key factors in population studies. *Journal of Animal Ecology*, **29**, 399–401.

Varley, G. C. and Gradwell, G. R. (1963a). The interpretation of insect population changes. *Proceedings of the Ceylon Association for the Advancement of Science*, **18**, 142–156.

Varley, G. C. and Gradwell, G. R. (1963b). Predatory insects as density dependent mortality factors. *Proceedings of the 16th International Congress of Zoology*, **1**, 240.

Varley, G. C. and Gradwell, G. R. (1968). Population models for the winter moth. In *Insect abundance*. Symposium of the Royal Entomological Society of London (ed. T. R. E. Southwood), pp. 132–142. Blackwell, Oxford.

Varley, G. C. and Gradwell, G. R. (1971). The use of models and life tables in assessing the role of natural enemies. In *Biological control* (ed. C. B. Huffaker), pp. 93–112. Plenum, New York.

Varley, G. C., Gradwell, G. R., and Hassell, M. P. (1973). *Insect population ecology, an analytical approach*. Blackwell Science, Oxford.

Vinson, S. B. (1972). Factors involved in successful attack on *Heliothis virescens* by the parasitoid *Cardiochiles nigriceps*. *Journal of Invertebrate Pathology*, **20**, 118–123.

Vinson, S. B. (1990). How parasitoids deal with the immune system of their host: an overview. *Archives of Insect Biochemistry and Physiology*, **13**, 63–81.

Visser, M. E. (1994). The importance of being large: the relationship between size and fitness in females of the parasitoid *Aphaereta minuta* (Hymenoptera: Braconidae). *Journal of Animal Ecology*, **6**, 963–978.

Visser, M. E. and Driessen, G. (1991). Indirect mutual interference in parasitoids. *Netherlands Journal of Zoology*, **41**, 214–227.

Visser, M. E., van Alphen, J. J. M., and Nell, H. W. (1990). Adaptive superparasitism and patch time allocation in solitary parasitoids: the influence of the number of parasitoids depleting a patch. *Behaviour*, **114**, 21–36.

Visser, M. E., Jones, T. H., and Driessen, G. (1999). Interference among insect parasitoids: a multi-patch experiment. *Journal of Animal Ecology*, **68**, 108–120.

Volterra, V. (1926). Variazioni e fluttuazioni del numero d'individui in specie animali conviventi. *Memorie della Accademia Nazionale dei Lincei*, **2**, 31–113.

Waage, J. K. (1979). Foraging for patchily distributed hosts by the parasitoid, *Nemeritis canescens. Journal of Animal Ecology*, **48**, 353–371.

Waage, J. K. (1983). Aggregation in field parasitoid populations: foraging time allocation by a population of *Diadegma* (Hymenoptera: Ichneumonidae). *Ecological Entomology*, **8**, 447–453.

Waage, J. K. and Lane, J. A. (1984). The reproductive strategy of a parasitic wasp. II. Sex allocation and local mate competition in *Trichogramma evanescens. Journal of Animal Ecology*, **53**, 417–426.

Waage, J. K. and Ng, S. M. (1984). The reproductive strategy of a parasitic wasp. I. Optimal progeny allocation in *Trichogramma evanescens*. *Journal of Animal Ecology*, **53**, 401–415.

Wahlberg, N., Moilanen, A., and Hanski, I. (1996). Predicting the occurrence of endangered species in fragmented landscapes. *Science*, **273**, 1536–1538.

Walde, S. J. (1991). Patch dynamics of a phytophagous mite population: effect of number of subpopulations. *Ecology*, **72**, 1591–1598.

Walde, S. J. (1994). Immigration and the dynamics of a predator prey interaction in biological control. *Journal of Animal Ecology*, **63**, 337–346.

Walde, S. J. (1995). Internal dynamics and metapopulations: experimental tests with predator–prey systems. In *Population dynamics* (ed. N. Cappuccino and P. W. Price), pp. 173–193. Academic Press, San Diego, CA.

Walde, S. J. and Murdoch, W. W. (1988). Spatial density dependence in parasitoids. *Annual Review of Entomology*, **33**, 441–466.

Walker, I. (1967). Effect of population density on the viability and fecundity of *Nasonia vitripennis* Walker (Hymenoptera: Pteromalidae). *Ecology*, **48**, 294–301.

Wang, Y. H. and Gutierrez, A. P. (1980). An assessment of the use of stability analyses in population ecology. *Journal of Animal Ecology*, **49**, 435–452.

Watkinson, A. R. (1990). The population dynamics of *Vulpia fasciculata*—a 9-year study. *Journal of Ecology*, **78**, 196–206.

Watson, A. (1971). Key factor analysis, density dependence and population limitation in red grouse. In *Dynamics of populations* (ed. P. J. den Bower and G. R. Gradwell), pp. 548–564. Centre for Agricultural Publishing and Documentation, Wageningen.

Watson, A. and Miller, G. R. (1970). Territory size and aggression in a fluctuating red grouse population. *Journal of Animal Ecology*, **40**, 367–383.

Watt, K. E. F. (1959). A mathematical model for the effect of densities of attacked and attacking species on the number attacked. *Canadian Entomologist*, **91**, 129–144.

Watt, K. E. F. (1965). Community stability and the strategy of biological control. *Canadian Entomologist*, **97**, 887–895.

Weisser, W. W. and Hassell, M. P. (1996). Animals 'on the move' stabilise host–parasitoid systems. *Proceedings of the Royal Society of London, B*, **263**, 749–754.

Weisser, W. W., Wilson, H. B., and Hassell, M. P. (1997). Interference among parasitoids: a clarifying note. *Oikos*, **79**, 173–178.

Wennergren, U., Ruckelshaus, M., and Kareiva, P. M. (1995). The promise and limitations of spatial models in conservation biology. *Oikos*, **74**, 349–356.

Werren, J. H. (1983). Sex ratio evolution under local mate competition in a parasitic wasp. *Evolution*, **37**, 116–124.

West, S. A., Flanagan, K. E., and Godfray, H. C. J. (1996). The relationship between parasitoid size and fitness in the field, a study of *Achrysocharoides zwoelferi* (Hymenoptera: Eulophidae). *Journal of Animal Ecology*, **65**, 631–639.

White, E. G. and Huffaker, C. B. (1969). Regulatory processes and population cyclicity in laboratory populations of *Anagasta kuhniella* (Zeller) (Lepidoptera: Phycitidae). II. Parasitism, predation, competition and protective cover. *Researches on Population Ecology*, **11**, 150–185.

Whittaker, P. L. (1984). The insect fauna of mistletoe (*Phoradendron tomentosum* Loranthaceae) in Southern Texas. *Southwestern Naturalist*, **29**, 435–444.

Wiens, J. A. (1989a). *The ecology of bird communities*. Cambridge University Press, Cambridge.

Wiens, J. A. (1989b). Spatial scaling in ecology. *Functional Ecology*, **3**, 385–397.

Williams, F. M. and Juliano, S. A. (1996). Functional responses revisited. *Environmental Entomology*, **25**, 549–550.

Wilson, H. B. and Hassell, M. P. (1997). Host–parasitoid spatial models: the interplay of demographic stochasticity and dynamics. *Proceedings of the Royal Society of London Series B—Biological Sciences*, **264**, 1189–1195.

Wilson, H. B., Hassell, M. P., and Godfray, H. C. J. (1996). Host–parasitoid food webs: dynamics, persistence and invasion. *American Naturalist*, **148**, 787–806.

Wilson, H. B., Godfray, H. C. J., Hassell, M. P., and Pacala, S. W. (1997). Deterministic and stochastic host–parasitoid dynamics in spatially extended systems. In *Modeling spatiotemporal dynamics in ecology* (ed. J. Bascompte and R. V. Sole), pp. 63–81. Springer, NY.

Wilson, H. B., Hassell, M. P., and Holt, R. D. (1998). Persistence and area effects in a stochastic tritrophic model. *American Naturalist*, **151**, 587–595.

Wilson, W. G., De Roos, A., and McCauley, E. (1993). Spatial instabilities within the diffusive Lotka–Volterra system: individual-based simulation results. *Theoretical Population Biology*, **43**, 91–127.

Woiwod, I. P. and Hanski, I. (1992). Patterns of density dependence in moths and aphids. *Journal of Animal Ecology*, **61**, 619–629.

Wolda, H. and Dennis, B. (1993). Density dependence tests, are they? *Oecologia*, **95**, 581–591.

Wolda, H., Dennis, B., and Taper, M. L. (1994). Density dependence tests, and largely futile comments—answers to Holyoak and Lawton (1993) and Hanski, Woiwod and Perry (1993). *Oecologia*, **98**, 229–234.

Wolfram, S. (1984). Cellular automata as models of complexity. *Nature, London*, **311**, 419–424.

Wylie, H. G. (1966). Some mechanisms that affect the sex ratio of *Nasonia vitripennis* (Walk.) (Hymenoptera: Pteromalidae) reared from superparasitized housefly pupae. *Canadian Entomologist*, **98**, 645–653.

Zwölfer, H. (1971). The structure and effect of parasite complexes attacking phytophagous host insects. In *Dynamics of populations* (ed. P. J. Den Boer and G. R. Gradwell), pp. 405–418. Centre for Agricultural Publishing and Documentation, Wageningen.

Index to Genera

Author Index

Subject Index